T0224958

Proportions and Their Music

Karlheinz Schüffler

Proportions and Their Music

What Fractions and Tone Sequences Have to Do with Each Other

Karlheinz Schüffler
Mathematik
Heinrich-Heine-Universität Düsseldorf
Düsseldorf, Germany

ISBN 978-3-662-65335-7   ISBN 978-3-662-65336-4   (eBook)
https://doi.org/10.1007/978-3-662-65336-4

This book is a translation of the original German edition "Proportionen und ihre Musik" by Karlheinz Schüffler, published by Springer-Verlag GmbH, DE in 2019. The translation was done with the help of an artificial intelligence machine translation tool. A subsequent human revision was done primarily in terms of content, so that the book will read stylistically differently from a conventional translation. Springer Nature works continuously to further the development of tools for the production of books and on the related technologies to support the authors.

This Springer imprint is published by the registered company Springer-Verlag GmbH, DE, part of Springer Nature.
The registered company address is: Heidelberger Platz 3, 14197 Berlin, Germany

If disposing of this product, please recycle the paper.

For my grandchildren
Anthea, Christian,
Helena and Hendrik

Foreword

finally he had recognized that not only in this modulatio (musica) the numbers rule, but they complete everything in the first place.

Augustine (de ordine XIV, From [5], p. 124)

There have always been numbers from which a mysterious magic emanated. Today they may be gigantic prime number monsters, which in their unimaginable size want to bring us closer to the unreachable edge of the comprehensible; in times of the awakening analysis, they were irrationalities and transcendences of famous numbers such as π and e as well as the kingdom of functions; in the many centuries of antiquity above all – chosen for higher things – they have been the counting numbers of our world of experience. Cultures and religions are connected with the numbers one, two and three; there is hardly a fairy tale without seven somehow playing its part, and the twelve apostles are certainly not the only example of the importance this number claims for itself. And even more specially the number six: it is the sum of its real divisors – and therefore perfect and the very first of its kind!

We can confidently assume that in former times there must have been a bickering among the numbers – certainly not without danger – about the undisputed sole significance. Reliable information about the outcome of this dispute is not known – but we have good reason to believe that some of them came to their senses and realized that together they were – eventually – capable of quite different things. The "1" had already recognized this in the beginning of its being: It connects the *rationes arithmetica* (1 : *n*) with the *rationes harmonia* (*n* : 1) since forever.

▶ So four numbers got together and decided to found the musica theoretica: $6 - 8 - 9 - 12$.

These four numbers were the first to recognize they could still guard a totally different perfection as a treasure than if they had remained alone: The two inner ones could call themselves the arithmetic and harmonic mean of the two outer ones. And they discovered exciting magic: When they revealed their proportions to Queen One and they switched from the arithmetic to the harmonic world, the inner ones among them reversed their roles as mean values: Arithmetic became harmonic, and harmonic became arithmetic. But what they still sheltered as a higher secret was their belief that they truly lived in just relations among themselves. Since being able to predict the future, they were certain of a number-nuisance called the geometric mean, which claimed the place between the two inner ones. A brave man from Gerasa – Nicomachus was his name – told them that just as the outer ones were in inverse proportion to this number-nuisance, so were the two inner ones too.

Envious of all this perfect symmetry, the "10" decided to at least occasionally ask to be admitted to this alliance, which was indeed granted to it from time to time and then with reservations – even if not by the great Pythagoras. But instead, other greats realized where its contribution to musica theoretica could ultimately lead.

Thus the Harmonia perfecta maxima 6 : 8 : 9 : 12 has arisen, and with the participation of the 10 – which was given the title of a contra-harmonic mean – or mediety – and which also managed to smuggle its front geometric proportional into the family as a little contra-arithmetic sister – the Harmonia perfecta maxima founded what was later called diatonic music of perfect octaves, fifths and thirds.

In this book, we will undertake a tour on the connection between the two cultural sciences of mathematics and music, which since forever and throughout the ages has led to everything that has become the basis of the musical theory of tones, intervals, sounds and scales. This is the theory of the

- **proportions and their consonance**,

which, in its millennia-long alliance, has produced the whole edifice of concepts and their living anchors among themselves.

The book is intended for both mathematicians and musicians, but above all for **readers interested in cultural history**. We also hope that **schools** will want to make use of the **helpful impulses** that we have incorporated into the conception of the text.

- For "**mathematicians**", the book offers a new, comprehensive presentation of the ancient theory of proportions, its axiomatics and the laws arising from it. At the same time, a new interesting application potential arises from the interconnection with music-theoretical examples.
- For the "**musicians**", the book offers the technical background from whose laws the theory of scales, chords and, above all, terminology can be understood in a sustainable way; in doing so, we also give a great deal of space to the concerns of the organ and its timbres born out of proportions. In this way, we recommend to **students** in particular that they try to access the fundamental theoretical concepts of their subject also through an approach guided by mathematics.
- For those **interested in cultural history**, we would like to give an insight into the vitality of the relationship between these two sciences, which has been and still is the focus of heated debate from antiquity to the present day.
- In particular, we recommend reading this book to those **teachers** who, for example, would like to work with their students respectively their pupils on interdisciplinary – and

therefore particularly acceptable and memorable – areas of application within the framework of separate advanced courses in their subject areas.

From a **mathematical point of view**, experience has shown that it is an enormous help to see that behind a "grey theory of calculation" there is a practice that can be understood and described precisely through this – in our case, the multitude of sound constructs.

Conversely, from a **musical point of view**, it is not really wrong to add mathematical elements to the musical levels of the scale theories, which not only serve the understanding excellently, but also – together with others – give the subject of "music theory", apart from its artistic enjoyment, the attitude of a science with all the characteristics that go with it.

▶ In short, this book is devoted to the mathematics of the ancient theory of proportions; we develop the subject matter from the principles and concerns of the also ancient music theory. We describe this theory and we use a colloquial language – very often and whenever possible – to explain our maths. Beyond that we choose all our application references and examples exclusively from the field of music and its theory of diatonic and non-diatonic structures and constructions. The guiding principle of our ideas is the "Harmonia Perfecta Maxima".

Introduction

The goddess Harmonia had been the daughter of Jupiter because of celestial harmony, and of Venus because of pleasure. . . .

Remigius (From [5], p. 129).

Since early antiquity, the two sciences of mathematics and music have belonged to the circle of "arts" – namely to the "Septem artes liberales":

- **geometria, arithmetica, astronomia and musica**

were what the ancient worldview regarded as "the sciences and their culture". But musica was not – as is commonly assumed – the art of making music: rather, it consisted in describing the regularities of **tones and their relationships to one another** with the aid of the laws of arithmetic of whole numbers. A very precise distinction was made between

- **musica theoretica and musica practica**;

and the first was considered a mathematical science; its object of theory were the consonances. These, however, were again defined as "proportions of monochordal divisions".

▶ For this reason, it is not surprising that, when viewed in the light of day, we encounter many basic concepts of music theory in a seemingly mathematical guise: "semitones, whole tones, octaves, sevenths, fifths, thirds, seconds, circle of fifths" – these are some of the essentials of the structural instruments of scale theory. We also encounter the word **"harmonic"** – at first seemingly separately – in both disciplines; however, it is actually anchored there in – several – profound units of meaning.

Strictly speaking, then the unique connection between music and mathematics has its early historical anchorage in the interplay between the **theory of the "consonantiae" and the theory of the "proportiones"**.

In fact, the historical "theory of music" – especially that of Greek antiquity – is an ongoing, continous discussion about consonances; it is just as finely detailed as it is, unfortunately, also inconsistent with many opposing terminologies and views. These consonances were – as mentioned – on the one hand and originally obtained as proportions of certain (permitted) ratios on stringed instruments, from which the Greek term *"chordōn symphōnia"* (interplay of musical instruments) originates. On the other hand, however, number games dominated ratios over the "musica practica".

This is mainly the "theory **of proportions**" as described by Aristotle (384–322 BC) and Euclid (330–275 BC). How closely this theory of proportions was connected with the science of music – i.e. the theory of consonantiae – can be seen from the fact that it originally started with Pythagoras as a purely musically motivated theory of intervals and scales; it then developed into a mathematical theory of number proportions and was finally developed in its most abstract form – in which even an abstraction from numbers to sets can be interpreted – essentially by Eudoxos (408–347 BC).

▶   It shows that the comparison of any "magnitudes" is based on a concept of proportions, which only knows the natural numbers and with them describes the commensurability of those magnitudes. Thereby, the core of the ancient "commensurability" consists in the assumption that two comparable things are always commensurable – this means: There is always one unit, so that both magnitudes are composed of this unit (as a whole number).

This assumption was – still at times of Pythagoras and long after – a law, and questioning this could end badly. For us it is certainly clear, that the demand for commensurability was equal to the sole existence of rational numbers – irrationalities were – so to say by decree – not existent. But the diagonal in the unit-square could have eliminated this error: Because there the length of the diagonal is $\sqrt{2}$ and this number cannot be written in the form n/m, the commensurability of $\sqrt{2}$ with 1 is not given. Characteristically, these numbers have always been called "not specifiable quantities".

The **science of music** thus corresponded to the science **of the proportions of tones** – the latter being quite free of a physical robe of existence. Thereby we can also use the analogy (equality) of

• "**Proportion of two tones**" and "**Interval**".
Thus, the Pythagorean (pure) fifth corresponds to the (ancient notated) proportion 2 : 3, the octave to the proportion 1 : 2 and the unison to the proportion 1 : 1. While Pythagoras created his music exclusively from these three building blocks by joining them together ("adjointing, iterating") and allowed nothing other than these three (prime) numbers 1, 2 and 3 and "derived numbers" from them, we encounter incredibly bizarre interval proportions in Greek tetrachordics with its doric, phrygian, (mixo-)lydian and many other scales. Thus, for example, the Phrygian scale of Archytas of Taranto (in the fourth century BC) knows the tiny enharmonion with the proportion 35 : 36 – i.e. of the rank of a very small quarter tone – as a "consonant" interval, apparently carrying the prime number 7 with it – one of many examples.

Both in the field of musical constructions and in the arithmetic of proportions itself, the "joining together" of proportions to form so-called

- **Proportion Chains**

plays an important role. This is how we derived – transferred to music – chords and scales. A particularly remarkable interweaving arises for these chains of proportions when, in addition, the classical mean values ("medieties") are brought into play: Already the kingdom of the universally known

- **Babylonian mean values (arithmetic – harmonic – geometric)**

offers all by itself a wealth of highly interesting symmetries between chains of proportions and their inversions, the "reciprocals", and this network of relationships can be described even more comprehensively if we take these and other – in antiquity considered to be a science on their own –

- **Medieties (mean values)**

into the consideration. To give a simple example:

▶   The arithmetically averaged proportion of the fifth $2 : 3 \cong 20 : 30$ yields the perfect **major** third $20 : 25 \cong 4 : 5$, the harmonically averaged proportion $20 : 30$, on the other hand, yields the perfect **minor** third $20 : 24 \cong 5 : 6$. Conclusion: major and minor of the just diatonic are born; the proportion chain of the diatonic minor triad is the "reciprocal" of the proportion chain of the major triad.

In its theory of proportions, antiquity knew about ten mean values – "medieties" – one suspects that a music-theoretical game with intervals, scales and their chords opens up in its own world of inexhaustible relationships.

To the Chapter Overview
In **Chap. 1 (Proportions)** we begin with some reflections on the origins of the connections between mathematics and music – as far as it is of importance in the context of the theory of proportions. Then we begin this theory based on historical concepts. Here we develop the set of rules of ancient arithmetic laws from a minimal stock of given "plausible" basic rules, to which the role of "axioms" can be assigned and which can be identified in historical descriptions. The chapter concludes with a section on proportion fusion – that means the layering of two ore more intervals to its sum or its "Adjunction", respectively "Fusion" – and we describe the calculus of proportion equations and its algebra.

Chapter 2 (Proportion Chains) establishes all the constructive elements of dealing with proportions and the necessary rules and justifying theorems. Proportion chains are inherently compositions of several proportions into a more complex construction. We then develop, thanks to a structure-giving algebra and by means of the mathematical notions like

- Similarity
- Reciprocity
- Symmetry
- Composition - the ability to join

an ordering system for such chains of proportions. A musical counterpart of these constructions are intervals, scales and their chords. And it is precisely this ordering algebra of proportions that is also responsible for corresponding applications in these musical areas. The interplay of symmetry and reciprocity is particularly attractive; in the proportion chain theorem, this algebra is shown comprehensively and in as general a form as possible.

   Chapter 3 (Medieties) is devoted to mean values. First, we present the three historically abstract and in the literature also readily available description possibilities for "mean value proportions", which we then combine by the "**Babylonian medietary trinity**"

- **geometric – arithmetic – harmonic**

with the theorems of **Nicomachus** and **Iamblichos**, which had acquired their perfect universal significance in the theory of music of that time with the description of the "**Harmonia perfecta maxima**" in the case of the Pythagorean octave canon, and we present the general version of these important principles of harmony and symmetry, which shows that all those formerly marvelled number games are general and natural.

   The effort to obtain chord and scale structures directly from averages of existing proportions led to further – nowadays largely unknown – medieties, such as the "contra-harmonic mean" or the "contra-arithmetic mean" or to "higher front, middle or back proportionals" and many other deeper mysteries. At the heart of this mathematics are the **theorem on the symmetry of the third proportional** and the **theorem on the symmetry of the classical means**, which describes the chains of proportions – usually tripartite – formed from these means in terms of their symmetry and similarity properties and their relationships to each other, as well as their calculation formulas. This is followed by a section characterizing the **geometric mediety** as the "power center" of all mean values. We present the main symmetry mechanisms of this undoubtedly most important mediety in a goal-oriented **theorem on the harmonia perfecta maxima abstracta**. This is followed by **the theorem on the harmonia perfecta maxima diatonica** of pure diatonics, in which we describe and prove all symmetries of general five-step musical proportion chains.

   In Chap. 4 (Proportion Sequences of Babylonian Medieties), we first discuss the basic possibilities of extending given proportions or chains to arithmetic, geometric or harmonic chains. This is now followed by a discussion of the **contra medieties sequences**, which can be defined as sequences of mean values to an exponential family of proportion parameters. The contra medieties show a deeper inner threefold structured symmetry, which is subordinated to the common (!) geometric mean, as long as one orders this infinite sequences of medieties accordingly.

▶   Here we have certainly entered uncharted territory, and the view of the musical
    interval ratios of all the infinitely many possible contra medieties described in some
    special literatures to biblical dimensions can certainly benefit from our tighter
    mathematical description.

The second focus consists of a presentation of the iteration procedure for sequences of
Babylonian medieties – with the result that we obtain two-sided pairs of proportion chains
with infinitely many members – with the remarkable observation, that their magnitudes all
lie on one curve – the **hyperbola of Archytas** – and furthermore continue the symmetries
of ordinary finite Babylonian proportion chains into their ramified infinities – a **Harmonia
perfecta infinita** is achieved – both for the sequences of Babylonian medieties and for the
sequences of contra medieties.

▶   This theorem on the "Harmonia perfecta infinita" embodies both an inspiring and
    unique combination of analysis and geometry on the one hand and ancient musicol-
    ogy and its modern theory on the other, and it represents – in some sense – the
    mathematical center of our text.

Finally, in the last **Chap. 5 (The Music of Proportions)**, we consider a variety of
applications of the symmetries found around proportion chains and their Harmonia perfecta
maxima in the musical empire of intervals, chords, and scales.

We start with the laws of string and tone – that indispensable instrument "**monochord**",
which represents the connection of *audible but not measurable with measurable but not
audible proportions.*

The theory of proportion chains then accompanies us into the presentation of some
musical tonal systems – such as the pythagorean, the diatonic and the so-called ekmelic
systems.

▶   We have placed great emphasis on the development of the **ancient intervals** and
    their laws from the rules of calculation of the theory of proportions, and in particular
    we offer a systematic discovery, acquaintance with and practice in the use of all the
    significant whole, half and quarter tones as well as the commas of harmony from the
    idea of proportions and its laws.

The connection between music and proportions is also often seen in chordal music: The
proportion chain 4 : 5 : 6 of the major triad yields in its reciprocal chain 10 : 12 : 15 – as
already mentioned – the minor triad; that here also the arithmetic mean 5 (of the numbers
4 and 6) changes to the harmonic mean 12 (of the numbers 10 and 15), however, no longer
surprises us after reading the central theorem about the symmetries of the **medieties** chains
as well as Nicomachus' theorem.

An extraordinary role in the construction of all scales, from the ancient Greek and ecclesiastical modes to the temperaments of the Bach era and finally to the simplified major and minor of our time, is played by the **tetrachords** those elements which characterize the fourth as a four-note scale composed of three steps. These structures can also be used to describe the architecture of **gregorian** and **ecclesiastical** tonal forms. The section **Modology** takes up this theme. A steep course accompanies the reader from the universal tonal system of antiquity, the "**systema teleion**" and its **octochords**, to the Greek and **ecclesiastical tonalities.**

And in the last section **(Proportions and the Organ)**, we explain the **arithmetic of the foot-number of stops ("Organ Stop Calculus")**, and this topic is typical of organ music and highly relevant to organists, including the phenomenon of acoustic "32-foot basses" and similar tricky physical-mathematical applications to acoustic proportions. As a result, it is more likely that we will gain a new understanding of the colorfulness of a given organ in its capacity as an orchestral instrument with its almost infinite sound possibilities.

▶   This arithmetic of the organ stops is also of an original nature and defined by the concept of proportions of intervals and combinations of sounds.

This special topic is rounded off by some examples from the exciting real world of the **dispositions** of this instrument.

In an **appendix**, there is also, among other things, a reference table of almost all the ancient intervals of the diatonic with their proportions, their frequency measures and their cent measures. We have also added those four functions whose graphic progressions offer both the mean value relationships and the symmetry laws of the Harmonia perfecta to the eye as memorable elements of the theory.

Finally Our exposition will make use of quite common forms of mathematical reading, which, apart from a way of writing and designating that has become customary, also consists in defining terms – especially the decisive ones – (reasonably) sharply wherever possible, and in presenting all facts as well as their internal connections, discoveries and conclusions in a familiar orderly system of definitions, propositions (most of which are called "**theorem**"), their proofs as well as helpful remarks and **musical examples**. In fact, we have deliberately avoided other applications of the theory of proportions, especially geometrically motivated ones, except in a single case. Thus, the reader will find suitable examples from the world of music theory not only in the last large Chap. 5, but also accompanying the entire reading.

▶   **Reading Suggestions**

We would also like to point out that this book tries to meet the different interests as well as knowledge and inclinations of its different readership: We can very well imagine that especially the fifth chapter is able to establish the entry as well as the usual start in the first chapter – and whoever then subsequently wants to get to the bottom of mathematical things will certainly soon be on the desired course of success by turning back the pages. Certainly, our diverse music-theoretical examples, which accompany the complete text, may also contribute to this.

For Use: Proportion Convention *a:b* or *b:a*?

What is an octave, what is a fifth?

Wikipedia says about the former one: "**Octave** (rarely *octav*, from **Latin** octava 'the eighth') is the term used in music for the **interval** between two notes whose **frequencies** act like 2 : 1."

In fact, this "ratio" has become the predominantly common answer to the question posed at the beginning – and in the case of the fifth, the indication of the ratio also follows 3 : 2. What is missing from the above definition, however, is an indication of which arrangement this ratio refers to. If we have two tones *a* and *b* that form an octave, which frequency ratio is then meant by 2 : 1 – *b* to *a* or *a* to *b* (*b/a* or *a/b*)?

Furthermore, contrary to the Wikipedia definition above, a musical interval [*a*, *b*] is not the "space between two tones", but the "ordered tone pair (*a*, *b*)" itself. Here *a* is a starting tone or reference tone and the second tone is the target tone.

Question How should we interpret a statement that an interval has the "ratio" $x : y$ – for example 1 : 2 or 2 : 1?

Anticipating later explanations, we find the following concept in the idea of proportions and its symbolic notation: If two quantities (magnitudes – for example, tones) *a* and *b* are given, then the proportion equation

$$a : b \cong 1 : 2$$

(read "a relates to b like (the number) 1 to (the number) 2") means, that the formative characteristic (pitch, frequency, length, content, weight...) of the magnitude *b* is twice as large as that of the magnitude *a*. And the equivalent fraction-arithmetically interpreted equation "*a/b* = 1/2" also has exactly the solution "*b* = 2*a*".

The question concerning the indication of proportions is thus the question of how tones are "measured". For this there are – in principle – the two forms:

A. The pitch proportion

Here we connect (but basically unconsciously) the frequencies of the tones (that is their fundamental physical vibrations). If we write the ratio of the vibrations of two ordered tones (*a*, *b*) as a proportion, it follows that the musical interval [*a*, *b*] has the proportion

$$a : b \cong \text{(frequency of the tone } a) : \text{(frequency of the tone } b)$$

and we speak of the pitch proportion. Here we interpret the sign $\cong$ as "corresponding to" – which means that the corresponding magnitudes are equal in the case of numbers – modulo a suitable multiplication with a common factor – in short, that the fraction (a/b) of the left side is equal to the fraction (frequency of the tone a /frequency of the tone b) of the right side.

→*The octave has the pitch proportion* 1 : 2.

B. **The monochord proportion (string/pipe length proportion)**

At all times when one knew nothing about vibrations and frequencies, tone relationships were usually measured by the length ratio of the vibrating monochord strings: If we have a tensioned string with the (fundamental) tone *a* and if we shorten this string by half, for example, the newer tone *b* of each of these halves will sound "an octave higher". The ratio of the lengths of the tone-producing strings L_a (fundamental) and L_b (octave tone) is therefore 2 : 1.

→*The octave has the monochord proportion* 2 : 1.

We have decided to use the pitch proportion (**A**) throughout this text. The proportions $(2 : 3), (3 : 4), (4 : 5)$ therefore stand for perfect fifths, fourths, major thirds and so on. Thus, if we say $a : b \cong 1 : 2$, we have described an (upward) octave.

There are several reasons for doing this, for one thing: By noting the proportion $a : b$ for a pair of notes, we think more musically than geometrically: our mind's eye is (always) accompanied by a keyboard that quite spontaneously and without being asked transforms the proportion into two notes and which we also imagine audibly in an inner world.

A second thought supports this choice in a quite different way: Harmonics – especially that of antiquity – is the science of tonal relations, which result from the proportions to the natural numbers. Thus we read, for example, in [Hans Kayser (10), p. 48 ff:]

> If I set the oscillation number of the fundamental equal to 1, then we look to see which tones come out for the oscillation numbers 2, 3, 4, 5, 6... these are then the octave above the fundamental then the fifth above this octave (2 : 3), then the double octave above the fundamental (1 : 4), then the pure (perfect) major third above the double octave (4 : 5) and then above this another minor third (5 : 6)...

From the so-called overtone spectrum of a tone, the proportions line up in the form $(1 : n)$ to the fundamental, and consequently the proportion representations of the intervals arise, which do not describe the relationship of the final tone to the reference tone, but follow the written direction of the proportion.

A third idea: harmony is concerned with far more than simple one-step proportions, but above all with multi-step chains. Thus the simple numerical proportion chain 4 : 5 : 6 stands for the "major chord of just diatonicism", or more precisely: three tones *a*, *b* and *c* form a major triad if the similarity equation

$$a : b : c \cong 4 : 5 : 6$$

is true. And just as you read from left to right, you get the proportions 4 : 5 for $a : b$, 5 : 6 for $b : c$ and 4 : 6 for $a : c$, that is, b has 5/4 times the frequency of a and c has 3/2 times it.

▶ Conclusion: By reading longer chains, the eye sees the sequence of intervals simultaneously in the sequence of proportions – without having to reverse the roles of end and start point.

Throughout the book, therefore, we use the (pitch) proportion measure 1 : 2 for the octave, and all intervals described by numerical proportions are to be interpreted accordingly.

The "frequency measure" b/a of a musical interval, on the other hand, remains unaffected: it is defined precisely by specifying the factor by which the frequency of a given (fundamental) tone a must be changed in order to produce the tone b – this factor is obviously b/a.

▶ Thus, the frequency measure of an interval $[a, b]$ focuses on the target note (b), whereas the proportion measure (A) focuses on the "ratio of the initial note to the final note".

From a superficial point of view, then, the frequency measure is apparently directly connected to the monochord proportion: Fifth 3 : 2 ↔ Frequency Measure 3/2.

Incidentally, both forms (A) and (B) are connected via the musical mechanism of action "up – down": The intervals $a : b$ to (A) and $a : b$ to (B) have the same frequency measure – but run in opposite directions. Thus, for example, the interval 2 : 3 to (A) is a perfect upward fifth, while the interval 2 : 3 to (B) describes a perfect downward fifth.

To illustrate all this in a final example, let us consider the chain of proportions – called the "**Senarius**", which was very famous in antiquity,

$$1 : 2 : 3 : 4 : 5 : 6.$$

We see that all number elements of their proportions come from the prime numbers (1), 2, 3 and 5, from which all intervals called "**emmelic**" arose. Intervals whose proportions also require other prime numbers besides 2, 3 and 5 as factors are, by the way, called "**ekmelic**" intervals. If we write notes over them, the notes played in just tuning at a given start c_0 (the "tonic") result $c_0 - c_1 - g_1 - c_2 - e_2 - g_2$, and we read off the step intervals from left to right (from note to next note) **and** concordantly to the chain of proportions:

Octave $(1:2)$ – Fifth $(2:3)$ – Fourth $(3:4)$ – major Third $(4:5)$ – minor Third $(5:6)$,

which corresponds to the score

We hope that this reading will serve to speed up the acquisition, because it is just perfectly aligned with the concept of proportions and their chains.

Acknowledgements

Düsseldorf and Willich, Germany

I would like to express my sincere thanks to Springer Spektrum and its chief editor, Ulrike Schmickler-Hirzebruch, and her colleague, Barbara Gerlach, for including the german edition of this book in their publishing program; I would also like to thank Dr. Annika Denkert and Stella Schmoll – both from Springer Spektrum: Annika Denkert has taken over the supervision of the publication of the german edition of this book from Ulrike Schmickler-Hirzebruch, and Stella Schmoll has thereby supervised the editorial matters up to the printing. I would also like to thank my former student, Mr. Sascha Keil, for his loyal cooperation and manifold help with technical realizations; my friend and colleague Stefan Ritter has rendered outstanding services with one or the other illustration, and I thank him warmly as well. In the last phases of the creation my wife saw me rather seldom; I was less driven by external deadlines – no, the deeper I immersed myself in these two interconnected ancient worlds, the more exciting I found the urge to fathom all these things – whereby it was also an urgent concern of mine to generalize and reprocess in our modern language ancient knowledge that can be found in hardly readable and understandable literatures and only with great effort, in order to give free rein to the urge to play, to conquer new things. Once again, I would like to thank Dr. Andreas Rüdinger and Dr. Annika Denkert from Springer Spektrum for the publication of the english translation of the german edition. I would also like to thank Dea Dubovci (University of Düsseldorf – Germany) for her critical review of the DeepL translation.

I thank my readers if they would like to accompany me on this path – lined by the science of mathematics – into the ancient world of music theory.

March 2019 and December 2023

Düsseldorf and Willich, Germany   Karlheinz Schüffler

Contents

Proportions

1

> Harmony consists of tones and intervals, and indeed the tone is
> the one and the same thing, the intervals are the otherness and the
> difference of the tones, and by mixing this, song and melody
> result. . . . *Aristoxenes (From [5], p. 79)*

There is no doubt about it: hardly any other term like that of "proportion" accompanies the
musical theory of intervals – thus also of tonal intervals – with all its meanings. Thus we
read everywhere that

> the interval of the pure fifth is in the ratio 2 : 3,
> the octave is defined by the condition 1 : 2

and so on. Almost everything connected with scales, their characteristic features, and
distinctions, is ubiquitously permeated by a language which – in wanting to give precise
descriptions – makes use of the terminology of "proportions" all around. In this chapter we
present a foundation of arithmetic with proportions. We begin with a classification of
various mathematical-musical concepts, oriented on some ancient ideas. Then we develop
the **notion of proportions** from these fundaments and introduce an **axiomatic of the laws
of arithmetic** – called the "theory of proportions" respectively the "calculus with
proportions" – which is centered mainly in Theorems 1.1 and 1.2. In doing so, we anchor
this theory of proportions on basic rules, which leads to the catalogue of ancient arithmetic
rules. Finally, we also adress the multiplication of proportions, viewing it as a process of
merging (**fusing**) two (or more) proportions - which musically corresponds to the layering
of intervals to form a new interval.

1.1   Arithmetica and Harmonia: The Genesis

Unfortunately, of Greek music – the oldest source of occidental culture – only theoretical material (writings) has survived – quite in contrast to architecture, sculpture, painting, coinage, gemology, and poetry. If there were only a single audio example – how much more revealing would such a stroke of luck be about the way music was understood. Certainly: our knowledge of ancient instruments allows us to guess at much of this and to recognize it as certain – but how it really felt in the ancient "musica practica" still gives us plenty of room for imaginative ideas and research. And a mere theoretical description of musical processes might be of very limited use to us. And yet: the more we know about the metrical data of the "musica theoretica", the more we can – also thanks to modern technology – listen into the sound structures of the "musica practica".

A chronologically ordered list of important older written literature of music-theoretical information contains many well-known names – like for example

- Aristotle (fourth century BC): 19th chapter of the "Problemata" and 5th chapter of the 8th book of the "Republica"
- Aristoxenes (pupil of Aristotle, end of fourth century BC): three books: Elements of Music
- Euclides (third century BC): Introduction to music as well as the division of the string
- Philodemus (first century BC): On music
- Plutarch (first century AD): Writing about music
- Aristides Quintilianus (second century AD): Work on music (in three volumes)
- Claudius Ptolemaeus (second century AD): Harmonics (three books)
- Julius Pollux (second century AD): Instrumentology
- Theo of Smyrna (second century AD): Mathematical description of music
- Nicomachus of Gerasa (60–120 AD): two books on music theory
- Anitius Manlius Severinus Boethius (sixth century AD): five books on music
- Michael Constantin Psellus (eleventh century AD): The Quadrivium
- Manuel Bryennius (fourteenth century AD): Harmonics

On the other hand, our current conception of music – namely as an art that reveals itself directly to acoustic aesthetics – was by no means valid in those times: music-making and the theory of music must be regarded as almost entirely separate areas. While making music on the instruments of the time, such as flutes, lyres and alike, was regarded as a low-level activity and "musicians" therefore tended to earn their bread at the lower end of a social order, music theory was part of the **"quadrivium"** – that supreme human science consisting of the four "mathematical" disciplines

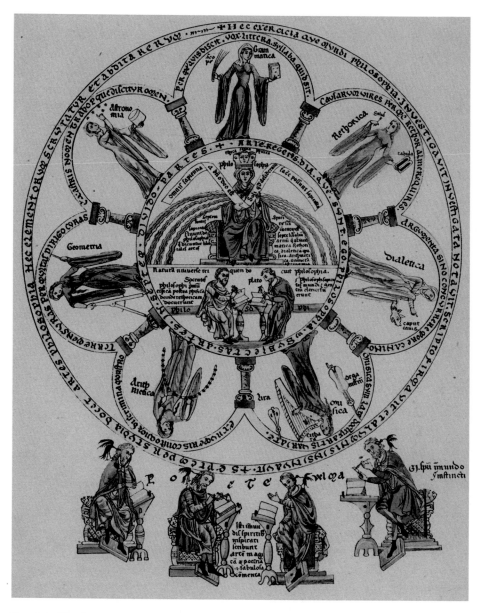

Fig. 1.1 The Septem Artes Liberales. (From the "Hortus deliciarum" of Herrad of Landsberg, twelfth century) (© akg-images/picture alliance)

- **Arithmetica – Geometria – Astronomia – Musica.**

These four liberal arts ranked even higher than those of the **"trivium,"** which consisted of the three philosophical arts

- **Dialectica – Grammatica – Rhetorica.**

Together they formed the **"Septem Artes liberales"**, as can be seen in Fig. 1.1 – the seven liberal arts, which until the beginning of the modern age – together with theology and medicine – defined what was called "science" and what was consequently the teaching at all historical universities. In the image of the four mathematical arts – i.e. the quadrivium – an inner structure emerges, which presented geometry with arithmetic and astronomy with music as closely connected:

- **Geometria ⇄ Arithmetica and Astronomia ⇄ Musica.**

It is precisely from the close relationship of these pairs of siblings that the richness of the common conceptual worlds is ultimately drawn. Indeed, in ancient writings we encounter a wealth of thought in which homage is paid to the significance of these connections with such unwavering devotion that we can (or could) gain a new approach to "scientific" thinking in ancient times. The following quotation from **Aristides Quintilianus** may serve as a sample, and we recall very briefly the Harmonia perfecta maxima already mentioned in the preface with its "sacred musical numbers" 6 – 8 – 9 – 12, to which the following refers:

> It is said of the musical numbers that they are all perfect numbers and sacred. This is especially true of the ration of the whole tone (the great second) 8 : 9. This numerical ratio expresses the harmony of the universe; for the planets are seven; the Zodiac adds an eighth sphere; thus the outstanding cosmic significance of the number nine is shown. If, however, 8 : 9 were substituted for 16 : 18, the number 17 would result in a very meaningful division of the whole tone into two semitones. But because the number 17 is the arithmetical middle between 16 and 18, so it indicates the original reason of the connection of the moon with the earth (etc.)... (From [6], p. 16).

The motor of this network of relations consisted in the fact that almost all forms of determining, describing and justifying made use of both language and the elements of the "theory of proportions", that form of mathematics of the time which pervaded both geometry and arithmetic in equal measure. If we think, for example, of the well-known **"ray theorems" of geometry**, we have a fairly good idea of how **thinking in "ratios"** came to be a formative form of the sciences of the Quadrivium. This thinking in ratios had its own laws together with their rules based on them – but a reliable and universally valid basis could not – how could it – really develop.

In the following sections of this first chapter, we will therefore introduce the calculus of the theory of proportions in more detail.

What Is the Theory of Proportions About?
In the theory of proportions "magnitudes" are compared and "set in relation to each other". Here two things are worth noting:

First: As already mentioned in the introduction, the notion of "commensurability" plays a role that should not be underestimated; it states that for two magnitudes a and b – if they are at all in proportion to each other (or: if they form a *ratio* at all) – one can say:

a relates to b as (the natural number) n relates to (the natural number) m.

If we write $a : b \cong n : m$, then the sign $\cong$ already expresses, that the area of the common number-calculation is not necessarily present here.

Secondly, that among the objects to be compared – i.e. those magnitudes which one brings into proportion with one another – all kinds of abstract things such as surfaces, bodies, sounds, angles etc. – and of course also numbers – are found, so that a symbol "$a : b$" would not necessarily be interpreted as a "fraction" obeying the simplest laws of calculation. In fact, it can be observed that very subtle descriptions of what

$$a : b \cong n : m$$

should be, in retrospect also lead to the surprising consequence that the commensurability requirement with its consequent rationality of all number relations can be replaced by a more universal – giving space to irrational numbers – level of argumentation: The limits of commensurability are thus exceeded after all – perhaps unconsciously – at least in one or the other case.

The theory of proportions now led directly to the **theory of medieties** and their abundance of "mean values" **(medieties),** which in turn determined the path of the "arithmetica" and the "harmonia" – which we will discuss later – towards the "musica". At first, these were the three Babylonian mean values, as they were presumably already known to the Pythagoreans; at least in late antiquity, Archytas ($\approx$430 – 350 BC) still lists the medieties that were also called **"Babylonian"**

- arithmetic mean,
- geometric mean,
- harmonic mean.

In antiquity, these were also known as **"musical medieties"**; we read, for example, in **Nicomachus of Gerasa**:

> ...*that the knowledge of the theory of the different kinds of medieties (first the arithmetical, second the geometrical, and third the inverse, which is also called harmonic) is exceedingly necessary for natural history and the theory of music, for the contemplation of the spheres (astronomy) as well as the laws of geometrical measurement in the planes (i.e. for all physics and branches of the quadrivium), but most of all for the understanding of the reading of the ancient scriptures...* (From [6], p. 118 et seq. 118 ff.)

To these classical medieties, which are also encountered in other ancient cultures such as the Egyptian and Persian, are added other medieties: first of all, these are

- the contra-harmonic mean and
- the contra-arithmetic mean,

whose role in pure diatonicism is quite decisive. Finally, some other mean values can be added; these are partly those that arise from iterated arithmetic and harmonic averages – i.e. proportions. In antiquity there were (broadly speaking) ten medieties.

In **music theory**, however, it is especially and almost exclusively these five aforementioned medieties that define the diatonic tonal space – with the geometric medieties only appearing in the background, but being in charge there – the later Sect. 3.5 gives ample justification.

The Theory of Music (Musica Theoretica)

The Theory of Music ("Musica Theoretica") now grew entirely out of the theory of proportions, namely by establishing music as an ordered cosmos of the basic building blocks **"intervals"**, whereby all this was expressed in the language of the theory of proportions.

▶ **Important**

A musical interval – as we would call it today – is an **ordered** pair of two tones *(a, b)*, and musica theoretica describes the interval as a proportion of these two tones (which we interpret, with our present knowledge, as the ratio of the vibrational frequencies (of *a* versus *b*)).

It is not the individual tone – its absolute "pitch", colour, frequency ("oscillation mass") and so on – that is decisive, but the ratio of those magnitudes of input (front member *a*) to output (back member *b*) of the interval in question.

Finally, intervals are joined together to form tetrachords and scales together with their chords: this is how the edifice of musica theoretica is built. If, on the other hand, we add proportions to one another, we create chains of proportions – the bridge from the tonal scales and their chords to the numerical relationships is built.

In these construction mechanisms from interval to scale, as in the theory of consonance, the calculus of medieties plays an outstandingly important role: in the most elementary context, it mediates the change from interval to scale

- **Major** and **minor** in the ordinary musical sense,
- **authentic** and **plagal** in the Gregorian context

as a number game, which is accompanied by astonishing symmetries. In the arithmetic of two intervals placed next to each other, it is above all the Babylonian mean values with their contra-partners which accomplished the construction of the Greek tetrachordic and the later diatonic.

All these medieties are defined as chains of proportions, and the musical interplay of their inner proportions with one another is most impressively discernible in the **harmonia perfecta maxima** – that "most perfect harmony":

▶ Since the Harmonia perfecta maxima draws its almost mystical power from the inner symmetry of arithmetic ("authentic") and harmonic ("plagal") chains of mean-value proportions, which in turn possess their symmetries primarily in the fact that they are reciprocal to one another and that the geometric medieties represents a centre of symmetry to all other medieties – irrespective of whether that mean value exists as a rational number or not.

We will not only encounter this Harmonia perfecta maxima very frequently in our text – rather, we have developed the description of ancient music theory, if it forms a scientific twin with the theory of proportions, from the idea of a **Harmonia perfecta universale.**

After this brief overview, we will now describe what can (but does not need to) be considered the basis of music-mathematical elements in the **linguistic, philosophical** and **arithmetical** sense. Here we can – understandably – only consider a very small part of the historical considerations – and even this only in the style of an overview. Interested readers will find a huge space in the vast literatures and will also encounter the most diverse views – may they appear scientific, speculative or even bizarre, strange in nature.

The Genesis: Arithmetica and Harmonia

The entire richness of the ancient intervals as well as the theory of mean values is revealed when the model of two reciprocal proportion sequences "Arithmetica" and "Harmonia" is taken as the basis for all considerations. This model is presented primarily in [6, 7], and the extremely demanding reading of these investigations, which date from the nineteenth century, develops over many hundred pages an unprecedented as well as admirable overall presentation of musical proportions. Accordingly, it is emphasized that the two sequences of proportions

$$1 : 1, \ 1 : 2, \ 1 : 3, \ \ldots \text{etc.} - \text{short } 1 : n \ (\text{Arithmetica})$$

$$1 : 1, \ 2 : 1, \ 3 : 1, \ \ldots \text{etc.} - \text{short } n : 1 \ (\text{Harmonia}),$$

which, in our present way of writing and looking at things, we rather associate with the two sequences

$$(n)_{n \in \mathbb{N}} = 1, \ 2, \ 3, \ 4, \ \ldots (\text{Arithmetica})$$

$$\left(\frac{1}{n}\right)_{n\in\mathbb{N}} = 1, \ \frac{1}{2}, \ \frac{1}{3}, \ \frac{1}{4}, \ \ldots \text{(Harmonia)}.$$

They form an (better: the) essential component of the ancient number-mystical harmonic symbolism and, as a consequence, are also to be counted among the anchors of the Aristotelian philosophy of that time.

Note These two forms of proportion are also a consequence of our way of seeing and notating we explained at the beginning of the book in the section "For Use":

A magnitude *(b)*, which enters into the proportion $a:b \cong 1:n$ with the magnitude $a \cong 1$, is identifiable with the natural number "n"; likewise, it follows from the ratio $a:b \cong n:$ $1 \cong 1:\frac{1}{n}$, that now the magnitude b can be regarded as a stem fraction "$\frac{1}{n}$".

And indeed the consequence

$$(n)_{n\in\mathbb{N}} = 1, \ 2, \ 3, \ 4, \ \ldots$$

is the prototype of **all arithmetic sequences,** and that of their reciprocal values, the "aliquot fraction sequence"

$$\left(\frac{1}{n}\right)_{n\in\mathbb{N}} = 1, \ \frac{1}{2}, \ \frac{1}{3}, \ \frac{1}{4}, \ \ldots,$$

has been called **"harmonic sequence"** since time immemorial. Certainly, the relationship with music can be considered as a source par excellence for this naming.

From the point of view of arithmetic – as well as from the point of view of the theory of proportions – these two sequences of proportions are both inverse and reciprocal to one another.

"Inverse": In musical terms, inverted proportions can be interpreted as opposing intervals: For example, if the proportion 1 : 2 means the interval of an upward octave to a given tonic, its inverse 2 : 1 is precisely the downward octave from that chosen tonic.

"Reciprocal" means that the step sequence of multi-membered proportion chains reverses its order when one changes from "arithmetica to harmonia" and vice versa – we describe this process in detail in Chap. 2.

The model of the two sequences Arithmetica and Harmonia, sketched in Fig. 1.2, thus generates the musica theoretica and is called the **harmonic-arithmetic proportion**

$$0 = (\infty : 1) \ldots (n : 1) \ldots (3 : 1) \ (2 : 1) \qquad (1 : 2) \ (1 : 3) \ldots (1 : n) \ldots (1 : \infty) = \infty$$

$$\textbf{HARMONIA} \qquad\qquad (1 : 1) \qquad\qquad \textbf{ARITHMETICA}$$

Fig. 1.2 Proportion model of Nicomachus

model – or also **the model of Nicomachus** (60–120 AD), who consistently "took it to its end". Certainly, however, it goes back via Aristotle and Plato to the Pythagoreans. Whereby however with Pythagoras himself only such proportions occurred, which could be derived from the prime numbers 2 and 3.

In this model, the "1" – represented as a proportion $1 : 1$ – has a special position that even reaches into theological dimensions; after all, it also unites arithmetica with harmonia. The model ultimately connects

> the unlimited divisibility of the unit up to the infinitely small with the quantity of the unit multipliable up to the infinitely large.
>
> Nicomachus; (see von Thimus [I, Appendix p. 12])

Strictly speaking, it is in this arithmetic of proportions that the inner connections of the similarities between Astronomia and Musica of the Platonic world view lie.

The extent to which this model also reaches into our present-day music theory may be illustrated by the following numerical game – anticipating later considerations:

Example

From the **arithmetic** sequence of proportions (here) the **major gender** arises: The first six members of the sequence of proportions (the **Senarius**) result in the chord

$$1 : 2 : 3 : 4 : 5 : 6 \quad \leftrightarrow \quad c_0 - c_1 - g_1 - c_2 - e_2 - g_2,$$

where the indices denote upward octave positions. The interval $c_2 - e_2$ of the proportion $4 : 5$ thus has the frequency measure 5/4 and is a pure major third, and the chord

$$4 : 5 : 6 \quad \leftrightarrow \quad c_2 - e_2 - g_2$$

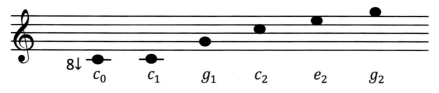

is therefore a C-major-triad in "pure diatonic" temperament.

The **harmonic** proportion sequence gives rise to the **minor gender.** The first six proportions now result in the chord read downwards

$$1 : \frac{1}{2} : \frac{1}{3} : \frac{1}{4} : \frac{1}{5} : \frac{1}{6} \quad \leftrightarrow \quad c_3 - c_2 - f_1 - c_1 - as_0 - f_0.$$

The interval $c_1 - as_0$ is in fact the major pure (downward) third $\frac{1}{4} : \frac{1}{5} \cong 5 : 4$, and likewise the interval $f_0 - as_0$ is a minor pure (upward) third in the frequency ratio $5 : 6$, and the upward read chord

$$f_0 - as_0 - c_1 \quad \leftrightarrow \quad \frac{1}{6} : \frac{1}{5} : \frac{1}{4} \quad \leftrightarrow \quad \frac{10}{60} : \frac{12}{60} : \frac{15}{60} \quad \leftrightarrow \quad 10 : 12 : 15$$

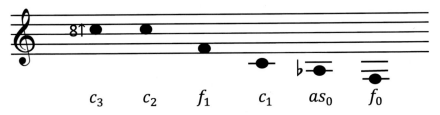

$$c_3 \qquad c_2 \qquad f_1 \qquad c_1 \qquad as_0 \qquad f_0$$

is the F minor triad in the also pure diatonic tuning. ◀

The Historical Number Families Perissos and Artios

The two families of proportions Arithmetica and Harmonia are, according to important researches, classified into

- **Perissos numbers (περισσος)** for the **arithmetic** proportion series.
- **Artios numbers (ἄρτιος)** for the **harmonic** proportion series.

although this assignment unfortunately is not interpreted uniformly. In many cases, this description has been interpreted as a decomposition of all numbers into the partitions **even/ odd,** which, however, could be incorrect because it would make no sense, according to the proponents and advocates of an arithmetic-harmonic number harmony: the partition of all "consonant proportions" into the two reciprocal proportion sequences "arithmetic" and "harmonic" is much more identifiable with the partition into Perissos and Artios numbers.

The terms "Perissos" and "Artios" can even be traced back to Pythagoras, although this requires a great deal of skill in dealing with ancient formulations – which is ultimately also where the scope for interpretation comes from. For example, Nicomachus gave the following description of these ancient Pythagorean terms:

> An artios number is that which takes the cut into the greatest and smallest according to the same; into the greatest: the as-large, into the smallest: the as-many, according to the natural interrelation of the two genera; a perissos number, however, is that which cannot bear this, but is divided into two unequal parts.

> von Thimus (I, appendix by R. Hasenclever, p. 12)

On the basis of this model of arithmetic-harmonic proportion sequences, we can now formulate the concept of consonance in what is probably its most original form:

> **Definition 1.1 (Consonance)**
> A (musical) interval – that is a proportion $\alpha : \omega$ – is called **antique-consonant** if this proportion is arithmetic or harmonic – or: if it is "derived" from this. The magnitudes α, ω are then to be regarded as natural (positive whole) numbers.
>   A proportion is called **derived** if it is a "difference-proportion" of two neighbouring arithmetic or harmonic proportions (this is explained in Sect. 1.4).

Thus a platform of historical mathematical music theory has been found: The detailed discussion of the two chains of proportions arithmetica and harmonia and their derivations allows – ultimately – for the

- Development of the structural theory of Greek-antique tetrachordics together with its later ecclesiastical tonal scale theory
- Anchoring of numerous names and terms of tetrachordics and ancient tonal systems.

The Models of Musica Theoretica
In the foreground of this ancient **musical theory of proportions** and its practical illustration are the following three models,

- **the monochord model,**
- **the canon,**
- **and the Tetractys,**

which establish the connection of music with the teaching of proportions both "visually" and "audibly".

On the **monochord** – that is, a taut string – pitches (intervals) are related to corresponding string lengths. In these experiments, tonal ratios are obtained by "lengthening" or "shortening" strings. Some names also derive from this monochord model, for example

$$\text{hemilion} \equiv 1 + \frac{1}{2} \text{ or epitriton} \equiv 1 + \frac{1}{3}.$$

For the string extended by half (hemilion), the old note (of the non-extended string) is a fifth above the new note (of the extended string), and for the string extended by a third (epitriton), this is a fourth. Both of these can be derived very quickly from the monochord rules of Theorem 5.1 in Chap. 5. We go into the laws of monochord in detail in this Chap. 5 (The Music of Proportions).

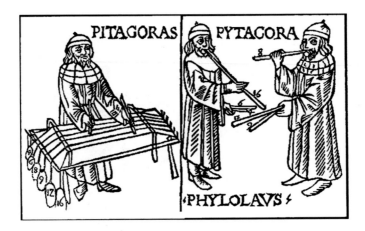

Fig. 1.3 The canon in pictures. (Source: Herrmann, The Ancient Mathematics)

The **canon** also basically consists of "one (but occasionally several) taut string laid across a ruler – divided into 12 parts", according to the description of an experiment by the late antique Gaudentius which he attributed to Pythagoras and which we see sketched in Fig. 1.3. We identify this (mostly) with the twelve-part number chain

$$1-2-3-4-5-\mathbf{6}-7-\mathbf{8}-\mathbf{9}-10-11-\mathbf{12},$$

from which one has obtained the most important musical proportions. The proportions 6 : 8, 6 : 9, 6 : 12, 8 : 9, 8 : 12 and 9 : 12 already contain all the constructive elements of Pythagorean music theory:

- 6 : 8 stands for the (Pythagorean) fourth, epitriton, diatesseron,
- 6 : 9 stands for the (Pythagorean) fifth, hemiolion, diapente,
- 6 : 12 stands for the octave, diplasion or diapasion,
- 8 : 9 stands for the (Pythagorean) whole tone, epogdoon, tonos.

So only the numbers 6, 8, 9, and 12 are needed for this.

As Fig. 1.3 shows us, there are various models of the lute canon; however, the weights shown there would never correspond to the imagined interval ratios, since the tensions and the weights are not in a "linear" relationship to each other: From Mersenne's frequency formula (see [16]) we see that the frequency of a tensioned string is not proportional to the tensile force P – but it is proportional to its root. Thus only a quadrupling of the weight leads to a doubling of the frequency – i.e. to the octave. Starting from the fundamental string with 6 weight units, the string of the fourth would then have to be loaded with 6∗

$(4/3)^2 = 32/3$ instead of 8, the string of the fifth with $6*(3/2)^2 = 27/2$ instead of 9 and the string of the octave with $6*(2)^2 = 24$ instead of 12 weight units.

Additionally to this oldest canon

$$6 - 8 - 9 - 12 \quad (\textbf{\textit{Pythagorean canon}})$$

then later came the so-called diatonic canon

$$6 - 8 - 9 - 10 - 12 \quad (\textbf{\textit{diatonic canon}})$$

which contains both the **pure major** third $8 : 10$ ($\cong 4 : 5$) and the **pure minor third** $10 : 12$ ($\cong 5 : 6$). It emerges from the Pythagorean canon through the addition of the contra-harmonic mediety (10) of 6 and 12, and if one still adds the partner "mirrored" on the geometric mediety – namely the contra-arithmetic mediety 72/10 – then one finally gains – with expansion by the factor 5 – the complete diatonic canon in whole-number magnitude form,

$$30 - 36 - 40 - 45 - 50 - 60 (\textbf{\textit{complete diatonic canon}})$$

Both new medieties are not derivable from the Pythagorean canon. Thus, it is precisely the major Pythagorean third (the ditonos) as the sum of two (Pythagorean) whole tones of the proportion $64 : 81$, which is distinguished from the perfect major third, which we now write in the form $4 : 5 \cong 64 : 80$, by a so-called "syntonic comma" $80 : 81$ – a scant quarter of a modern semitone step. In this by no means tiny difference lies – to put it briefly – also exactly the inner cause for the theory of tone scales, their contradictions including deficits and their numerous forms – from the mean tone to the modern equal temperament (cf. [16]).

Finally, the **Tetractys** is undoubtedly the *most mystical* model of musical Pythagorean number symbolism: the **Pythagorean fundamental equation of music**

$$\text{Octave} = \text{Fifth} + \text{Fourth}$$

is symbolically and memorably connected with the number system.

The Tetractys – sketched in Fig. 1.4 – creates, besides the connection to music, also several members of its basic numbers to the mathematics of the world of experience:

- **one** dot stands for the **"point"**,
- **two** points determine a **"distance"**,
- **three** points define the triangle, that is, the **"plane, area"**,
- **four** points form a tetrahedron, thus a **"space"**.

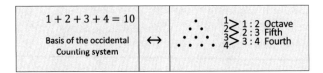

Fig. 1.4 The musical range of the Tetractys

Thus the dimensions of geometry and its objects are symbolized by the tetracty pyramid. Finally, religious, ideological references are also associated with the tetractys – but we do not wish to go into these here.

Now we turn to the actual theory of proportions.

1.2   Proportion and Similarity Equivalence

The "theory of proportions" can be divided into three subdivided phases:

1. The **musical theory of proportions.** It is the oldest, dates from the era of Pythagoras (or earlier) and was described and handed down by Euclid – and later by Ptolemaeus. In this theory, the intervals and their relationships to one another are described by means of the tetractys and, above all, by means of the canon.
2. This musical theory of proportions was later generalized and appears as a **theory of proportions of the integers** (i.e. of the natural numbers). This is described in books VII and VIII of Euclid's *Elements,* and it could also be called an "algebra of commensurabilities" – that is, fractional calculus.
3. The most general level, however, is reached by the **theory of proportions of Eudoxos,** in which not only numbers, but also more general **"magnitudes"** (megethē, horoi, magnitudes, quantities) stand in "proportion" to each other. Here, the concept of proportion even goes beyond "commensurability" – that is, beyond rationality: even incommensurable magnitudes are treated in the theory of proportions, which – unspoken – represents a certain preliminary stage to irrationalities.

When we see a ratio written today – for example, a "ratio" $2 : 3$ – we quite naturally make the associations

$$2 : 3 \rightleftarrows \frac{3}{2} \text{ or } \frac{2}{3} \text{ respectively } 2 : 3 \rightleftarrows 1.5 \text{ or } 0.\bar{6}$$

(depending on the reference point of the proportion; see the section "For use" at the beginning of the book), and a fortiori, in the case of a more general "proportion equation"

$$a : b = \alpha : \beta - \text{ or better still } a : b \cong \alpha : \beta \quad -$$

for two things (magnitudes) a and b and with – mostly natural – numbers α and β quite different notions are possible, and obviously there is a clear room for interpretation. Thereby it quickly becomes clear, that the above three expressions have quite their own meanings and fields of application:

- Through the symbolism as a **proportion "2 : 3"** it becomes very clear that here "two things are compared", no matter on the basis of which specific measurable characteristics this has to be done.
- The fraction "2/3" (or its inverse fraction "3/2"), thanks to its numerator-denominator geometry, still gives information that there is an "origin" of this expression; however, it immediately assigns this symbol to the calculus of fraction algebra and in this way loses something of the typical character of a proportion: both the comparison and the relational role of the magnitudes.
- The value "1.5" (as well as its reciprocal value $0.\overline{6}$) has now completely lost the character of a comparison; on the contrary, its role is to be seen in entrusting the proportion to a numerical yardstick.

Consequently – at least in the area of our music – the strengths

- of **proportion** are in clarity with regard to the structural arrangement ("who with whom"), their formative musical terms and their interrelationships with one another,
- of **fractional arithmetic** are in the equally exact and fast calculus in the set of rules for calculating with musical values,
- of **numerics** are in areas where numerical data comparisons are involved, as well as in areas where commensurability – and thus also fractional arithmetic – must be abandoned, as for example in all equal-step temperings.

Of course, the advantage of these different considerations is obvious, which is that we can – depending on the reference and at least nowadays – change the modes of representation, for example when comparing proportions. And that they can be compared – this too is a cinch if we use the algebraic or the numerical reinterpretations (assuming that the fractional calculus would be known again).

Let us bear in mind, however, that much of this – fractions, decimals and the like – did not exist in this conciseness in antiquity, if at all. An impressive example of this is provided by an introductory account of the ancient theory of proportions, such as that handed down by Eudoxos (408–375 BC) and others, and as we find it in Book V of Euclid's *Elements* (330–275 BC) (cf. [17], B. L. van der Waerden, p. 89 ff.).

Unfortunately, we do not find in Euclid himself any direct definition explaining **what** actually is the ratio of two quantities, nor what actually are **"quantities"** (magnitudes, horoi); however, we take up a formulation of Eudoxos to formulate a first "definition":

Definition 1.2 (Proportion _a_ : _b_)
"Two **quantities** _a_ and _b_ (**magnitudes**) stand in a relation (**proportion, ratio, analogon**) if a multiple of one exceeds the other and vice versa."

If two quantities _a_ and _b_ form a proportion, this is expressed by the symbol _a_ : _b_. The symbols _a, b_ themselves – as components of this proportion – are labelled with the Latin word magnitudes, but they are occasionally also designated by the Greek **"horoi".** Today one would say:

$$\text{the magnitudes } a \text{ and } b \text{ form a proportion } a{:}b$$
$$\Leftrightarrow \text{ there exists } n, m \in \mathbb{N}, \text{ such that } na \geq b \text{ and } mb \geq a,$$

where the symbol "$\geq$" is to be interpreted appropriately (for example, by "comprise"), and it is understood by the inequality that one quantity comprises the other. Similarly, intuitively, the multiplication "_na_" is explained by "multiple (_n_-fold) of magnitude _a_" and is not further explicated. If the horoi _a_ and _b_ are – as very often – just ordinary numbers, the analogon _a_ : _b_ is called a (mathematical) **logos.** If on the other hand _a_ : _b_ is a **musical** proportion (an interval, and the horoi would then be measures of relative or absolute pitches) this is also called **diastēma.**

Comments

1. These definitions obviously go beyond pure numerical quantities in two respects: On the one hand it concerns the generality of magnitudes: For example, geometric magnitudes (distances, areas, bodies) can be "compared" in this way. The expressions "_na_" are then to be understood in such a way that _n_ copies of the object _a_ are summarized. We recognize, on the other hand, that in this way quantities which stand in an "irrational" relation to each other can also have a "ratio" to each other. For example, in a square the comparison

$$\text{diagonal } (d) \geq \text{side } (s) \text{ and 2 sides } (2s) \geq \text{diagonal } (d)$$

   is valid.

2. It may seem a bit "fussy" and serve the reputation of mathematicians to want to be more precise than exact in a questionable way: Nevertheless, already in this definition we see how the term _a_ : _b_ is actually to be classified:

   ▶   _The symbol a : b stands a priori **not** for an expression to be calculated, but for a **"mathematical statement"** – namely that two things (a and b) stand in "relation" to each other in the first place, "they form a ratio"._

Finally this relation might be embodied by a number, nevertheless an "equation" like e.g.

$$a : b = 3/2 \text{ or } a : b = 1,5$$

does not make sense, because "$a : b$" first of all does **not** have to be a number which is supposed to be equal to another. This is true even in the case of numbers – magnitudes: Thus 3 : 2 is a proportion and not an actual division "3 divided by 2" or vice versa.

This explains the different approach mentioned at the beginning, which is reflected in the three forms of the

- Proportion – as a mathematical and calculable statement,
- Proportion – as fractional arithmetic and "division process",
- Proportion – as numerics and mere "numerical value indication"
  can be interpreted.

Thus, while the meaning of the symbol $a : b$ can generally only be recognized "intuitively" despite the above determination, the question of when three, four, or more horoi are *analogon* is now described by determining when two analoga are similar – or even "equal". In Book V of the *Elements* Euclid formulates this in this way:

> **Definition 1.3 (Similarity of Proportions and Magnitudes)**
> Four quantities a and b as well as c and d may be **analogous in pairs,** e.g. form the proportions $a : b$ and $c : d$.
>   Then we call the two **proportions** $a : b$ and $c : d$ **similar** (or **equal**) and then write $a : b \cong c : d \Leftrightarrow$ each of the following three conditions is true:
>
>   (a) If for two numbers $n,\ m \in \mathbb{N}$ $na > mb$ is valid, then $nc > md$ is also valid.
>   (b) If for two numbers $n,\ m \in \mathbb{N}$ $na = md$ is valid, then $nc = md$ is also valid.
>   (c) If the relation $na < mb$ is valid for two numbers $n,\ m \in \mathbb{N}$, then $nc < md$ is also valid.
>
> The quantities (horoi) are then also called **proportional** or **proportionate.**

If now in the case of numbers the proportion $a : b$ corresponds to the division a/b, then instead of "similar" the word "equal" is appropriate, and one immediately practices the abbreviated description

$$a : b \cong c : d \Leftrightarrow a : b = c : d \Leftrightarrow ad = bc.$$

In general, then the proof of the similarity or equality of two proportions consists in the comparatively tedious discussion of all three cases described in the definition, where we note that all expressions of the form *"na"* *have* the already mentioned "set-theoretic" meaning of a (whole-number) multiplication of the magnitude *a*.

Both the concept of "proportion" and that of "similarity", which obviously represents a step into a "theory of equations", simultaneously allow and require a first platform of theoretical-computational rules. It may well be, that there are sources of reasoning going further back – nevertheless we grant these basic rules the status of "axioms", which means, that we will not or can not prove them. We will describe this in the following Theorem 1.1:

Theorem 1.1 (The Basic Axiomatic Rules of Similarity)
1. **Basic rule:** The property of forming a **proportion** satisfies the following laws:
    1. **Reflexivity:** For all magnitudes *a*, $a : a$ is valid.
    2. **Symmetry:** If *a* and *b* form a proportion, so do *b* and *a*, their **inversion,**
       $$a : b \Leftrightarrow b : a.$$
    3. **Transitivity:** If *a* and *b* as well as *b* and *c* each form a proportion, then *a* and *c* are also in proportion, symbolically:

$$a : b \text{ and } b : c \Rightarrow a : c.$$

The **proportions** thus form an **"equivalence relation"** on the set of all pairs of magnitudes (more precisely: on suitable subsets of all pairs of magnitudes), since they possess the three properties characteristic of this: reflexivity, symmetry and transitivity. A further consequence is that every proportion has an **inverse** proportion (which is also unique only up to similarities):

$$A \cong a : b \Leftrightarrow A^{\text{inv}} \cong b : a \text{ is called an } \textbf{inverse proportion} \text{ to A.}$$

2. **Basic rule:** The term **"similarity"** also has the basic structures of an "equivalence relation"; this is expressed in the three laws of calculation
    1. **Reflexivity:** $a : b \cong a : b$,
    2. **Symmetry:** $a : b \cong c : d \Leftrightarrow c : d \cong a : b$,
    3. **Transitivity:** $a : b \cong c : d \text{ and } c : d \cong e : f \Rightarrow a : b \cong e : f$
    which can be read off from the above definition and which are valid for all proportions (and are usually also applied tacitly without any special mention).
3. **Basic rule:** For all magnitudes *a* and all natural numbers *n* we have the similarity:
    Norm: $a : a \cong n : n \cong 1 : 1$.

The concept of proportion is now profoundly connected to music by means of the basic musical structure of the interval, and we will explain this in the following example:

Example 1.1 Proportions and Musical Intervals

In modern terms, a musical interval I is the set of all ordered pairs of tones for which the ratio of the two fundamental frequencies f_1, f_2 of its two tones is always the same; and then the quotient f_2/f_1 is its **frequency measure** |I|.

In the context of historical music theory, which is largely based only on integer ratios, intervals are understood as proportions $a : b$ which are "similar" to the (true) frequency ratio. $f_1 : f_2$

Historically, the concept of frequency has been ignored, and we find only a classification of integer proportions as musical intervals (which at first seems manageable). The most important of these are:

the unison (1 : 1),
the perfect (pure, just or also pythagorean) fifth (2 : 3),
the octave (1 : 2),
the major (pythagorean) whole tone "Tonos", (8 : 9)
the (smaller) pythagorean semitone "Limma" (243 : 256),
the syntonic comma (80 : 81),
the pythagorean comma (524.288 : 531.441).

A comprehensive list of predominantly diatonic intervals can be found in the appendix. ◄

Now, in the ancient theory of proportions, a concept runs through – almost invisibly – almost all the concepts of mathematics of that time: this is the concept of **"commensurability"**.

What is understood by it? Well, it was the **pythagorean doctrine,** considered sacred, which expresses – in short -,

that all comparable magnitudes can also be compared in integer terms, in short: if a and b are in proportion, then there should be integers (positive) n and m such that the magnitude m∗a corresponds to the magnitude n∗b (see [3], p. 49 ff.).

That this cannot be true in general – due to the far-reaching notion of proportions of Eudoxos – is obvious. Therefore we devote an adapted definition to the so-called pythagorean doctrine:

Definition 1.4 (Commensurable Proportions)
Two magnitudes a and b are called **commensurable** if there are two *natural* numbers n and m such that a proportion of the form

$$a : b \cong n : m$$

exists – casually formulated: $ma = nb$; whereas the equality relation does not always make sense – in contrast to the proportion relation. Thanks to the "cross rule" as well as the "multiplication rule", which we present in the following theorem, this relation is also similar to the proportion $a : n \cong b : m$ and this in turn to the proportion $ma : nb \cong 1 : 1$, which is why these formulations could serve as definitions as well.

Comment
This pythagorean doctrine can also be interpreted in the context of the discussed possibilities of interpretation of the symbol $a : b$ in such a way that **all** proportions $a : b$ (must) have the proportion form respectively the fractional arithmetic form

$$a : b \cong n : m \text{ – respectively } - a : b \text{ corresponds to a fraction } a/b = n/m$$

and thus be characterized by a rational number.

▶   *However, we have known for a long time that this wishful thinking is just as wrong as it has its limits: The diagonal in a square has a non-rational relation to its sides ($\sqrt{2}$), and the pythagorean symbol, the regular 5-corner, has irrational relations, not to mention the relation of circles to radii and squares.*

We note – using the rules mentioned in Sect. 1.3 below – that, however, at least the equivalence

$$(m * a) : (n * b) \cong 1 : 1 \Longleftrightarrow a : b \cong n : m$$

is proved by the multiplication rule together with the cross rule.
In the next Sect. 1.3 tangible laws of calculation are waiting to be discovered.

1.3   The Calculation Laws of the Theory of Proportions

The Definitions (1.2), (1.3), and (1.4) presented in the previous Sect. 1.2, as well as the Definition (1.5) that follows later, are the most important conceptual elements of the entire theory of proportions, which then consists of finding – and also proving – a colourful palette full of regularities. In the following theorem, we list the most important of these rules, using a modern form of expression; samples of ancient texts are appended.

▶   **Important**
It is also interesting to note that only two or three main rules suffice to construct this ancient arithmetic about proportions: These are

- the cross-rule,
- the sum-rule,
- the difference-rule.

whereas the latter could still be obtained from the first two – and almost all arithmetic operations of the theory of proportions can be obtained by means of these building blocks.

Theorem 1.2 (The Axiomatic Calculation Rules of the Theory of Proportions)
1. **The cross-rule:** If the proportions are similar $a : b \cong c : d$, both the inner members (b and c) and the outer members (a and d) can be interchanged:

$$a : b \cong c : d \Leftrightarrow a : c \cong b : d \Leftrightarrow d : b \cong c : a.$$

2. **The sum-rule:** For proportions $a : b$ and $c : d$, where the magnitudes a and c on the one hand and b and d on the other hand can be "added" or joined together to form a "whole" – we write $g := a + c$ and $h := b + d$ – applies:

$$a : b \cong c : d \Rightarrow a : b \cong (a + c) : (b + d) \equiv g : h.$$

Generalization: This rule can be directly generalized to several proportions (with addable magnitudes a_k and b_k):

$$a_1 : b_1 \cong a_2 : b_2 \cong \ldots \cong a_n : b_n$$
$$\Rightarrow (a_1 + \ldots + a_n) : (b_1 + \ldots + b_n) \cong a_1 : b_1.$$

(continued)

Theorem 1.2 (continued)

   3. **The difference-rule:** For proportions $a : b$ and $g : h$, where a *is* a "part" (or: "piece") of g and b is a "part" of h – which we also write as $g = a + c$ and $h = b + d$ and where then $c := g - a$ and $d := h - b$ is called the "rest" of a in g as well as d is called the "rest" of b in h – applies:

$$a : b \cong g : h \Rightarrow a : b \cong (g - a) : (h - b) \equiv c : d.$$

   Remark: (The equivalence of sum and difference rule)

   In both the sum rule and the difference rule not only the implication "$\Rightarrow$" holds, but also the full equivalence "$\Leftrightarrow$" holds, because these two rules alternately cause the inversion "$\Leftarrow$" of the other rule. Thus both are equivalent.

We also understand these basic rules as **"axioms"**, although one or the other "proof" can be found in the ancient writings. In the first place, almost exclusively the rather complicated definitional anchoring of the term "similarity" (Definition 1.3) comes into play here – a testimony to the exceedingly astute argumentations of the ancient sources around Eudoxos and Euclid, if one likes to study the historical treatises. However, considering some vague anchors of the terms and their historical usage, exact proofs are very problematic anyway – but this leaves no doubt about a solid foundation of the set of rules. Using our fractional arithmetic of today, these rules would be evident by simple(st) calculations – if one would restrict oneself to mere "numerical relations". Nevertheless two remarks to the justifications of Theorem 1.2 may be mentioned:

1. In the case of the cross-rule (1) it would be sufficient to demand and prove only the interchangeability of the inner members (inner application) – then that of the outer members would result for reasons of symmetry: Since

$$a : c \cong b : d \Leftrightarrow b : d \cong a : c$$

   holds, outer members become inner members and vice versa. Otherwise, however, we consider the cross-rule axiomatically.

2. It is very easy to view the difference-rule as a consequence of the sum rule. For example, from the relation

$$a : b \cong (g - a) : (h - b)$$

   follows immediately the relation $a : b \cong g : h$ by adding the left to the right side.
   From these basic rules we now obtain the "application rules" derived from this:

Theorem 1.3 (The Rules of Application of the Theory of Proportions)
Application rule 1 (Reverse rule): $a : b \cong c : d \Leftrightarrow b : a \cong d : c$
Application rule 2 (exchange rule/substitution rule/truncation rule): If for two
   magnitudes a and $\tilde{a}$ their proportions to another magnitude b are similar, then their
   proportions to all magnitudes (which are in proportion to them) are similar:

   (a) **Exchange rule:** $a : b \cong \tilde{a} : b \Leftrightarrow a : c \cong \tilde{a} : c$ for all magnitudes c
         (with $a : c$ or $\tilde{a} : c$). This results in two ways of reading, which we
         define as rules:
   (b) **Substitution rule:** $a : 1 \cong \tilde{a} : 1 \Rightarrow a : b \cong \tilde{a} : b$ for all magnitudes b,
         which are in proportion with a and $\tilde{a}$. And the inversion leads to the
   (c) **Cancelling rule:** $a : b \cong \tilde{a} : b \Rightarrow a : 1 \cong \tilde{a} : 1$.
        Frequent application: Let a similarity relation $a : b \cong c : d$ apply. If then $\tilde{a}$ is
        a magnitude for which $a : \tilde{a} \cong 1 : 1$ is, then a can be substituted –
        i.e. exchanged – by $\tilde{a}$ while preserving the similarity, symbolically expressed
        in this way:
   (d) **Exchange rule:** $a : \tilde{a} \cong 1 : 1 \Rightarrow (a : b \cong c : d) \Leftrightarrow (\tilde{a} : b \cong c : d)$.

Application rule 3 (multiplication rules)
If we use the symbol $n * a$ to denote the n-fold combination of a magnitude a, or the
n-fold enlargement (multiplication), these rules apply:

   (a) For all proportions $a : b$ is $a : b \cong (n * a) : (n * b)$ for each $n \in \mathbb{N}$
   (b) If n, $m \in \mathbb{N}$ are arbitrary "multiples", the following applies

$$a : b \cong c : d \Leftrightarrow (n * a) : (m * b) \cong (n * c) : (m * d).$$

Application rule 4 (commensurability)
The property of commensurability carries over in several ways:

   (a) **Invariance under similarity:**
        If $a : b$ is commensurable and if $c : d \cong a : b$, so is $c : d$ commensurable.
   (b) **Commensurability is an equivalence relation:**
      (i) For all magnitudes a, $a : a$ ($\cong 1 : 1$) is commensurable (reflexivity).
      (ii) If $a : b$ is commensurable, so is $b : a$ (symmetry).
      (iii) If $a : b$ and $b : c$ are commensurable, so is $a : c$ (transitivity).

(continued)

Theorem 1.3 (continued)
Application rule 5 (The "Piece formula(s)")
If X and Y on the one hand, and α and β on the other, are each addable magnitudes
(where α and β may be positive numbers), we have the following three mutually
equivalent forms of mutually similar proportions:

(a) "Piece to the rest" form: $X : Y \cong \alpha : \beta$.
(b) "Piece to piece" form: $X : \alpha \cong Y : \beta$.
(c) "Piece to the whole" form: $X : (X + Y) \cong \alpha : (\alpha + \beta)$.

Before we justify this list of all the most common laws of arithmetic from the basic rules,
let us first list some comments.

1. Historical Interpretations of Piece Formulas:
    If one designates with $X + Y$ the "whole", then X is a "piece" of the whole and $Y =$
$(X + Y) - X$ is then the so-called "rest", and likewise one interprets $\alpha + \beta$, α and β. The
proportion $X : Y$ then expresses the ratio of **"piece to rest"**, the proportion $X : \alpha$ the
"piece to piece" ratio and the proportion $X : (X + Y)$ finally that of **"piece to whole"** of
the two horoi families X, Y and α, β.
2. Language Descriptions:
    Usually one finds textual interpretations of all these rules and terms in the antique
writings, whose transfer into an understandable colloquial language seems not seldom
quite adventurous; at least very many researchers of those writings state that with this or
that term a due portion of fantasy must flow in, if one wants to understand useful rules.
Such textual – but well readable – descriptions would be about these:
    1. *"If four sizes are proportionally equal, then the proportion of the forelimbs is equal
       to that of the hindlimbs"* (cross-rule).
    2. *"If two proportions are equal, so are their inverses"* (inversion-rule).
    3. *"Two quantities that have the same relation to a third are (proportionally) equal to
       each other"* (exchange rule).
    4. *"If any number of quantities are proportionally equal, then all the forelimbs together
       must relate to all the hindlimbs together as the single forelimb relates to the
       corresponding single hindlimb."*
        This is the general sum-rule; the condition of an addition-possibility is not
    explicitly mentioned – but it is assumed, because otherwise the rule could not be
    interpreted sensibly.
    5. *"A piece relates to the piece as the whole relates to the whole, then the rest must also
       relate to the rest as the piece relates to the piece or as the whole relates to the whole.*
       (This is the difference-rule, respectively one of the change formulas)".

3. Sum- and difference-rule are – thanks to the cross-rule – also applicable, if in the similarity-proportion $a : b \cong c : d$ not the magnitudes a and c as well as b and d are addable respectively subtractable, but if instead this is valid for a and b respectively as well as for c and d: Then the cross-rule leads by means of

$$a : b \cong c : d \Leftrightarrow a : c \cong b : d \cong (a \pm b) : (c \pm d)$$

to the possible applications of these rules.

4. The symbols $a + b$ or $a - b$ encountered in the listing of rules (2) and (3) for the formation of new magnitudes are, of course, to be questioned for each individual case with regard to their sense of purpose; the same applies to "equality". To give an – amusing – example:

   If a six-carton of French red wine (W) costs €30 (P), we could divide that in proportion calculus by

$$W : P \cong 1 : 1$$

   but this does not mean that $W = P$, and neither would the summation formally possible via the cross-rule $W : 1 \cong P : 1$ being

$$(W + P) : 2 \cong W : 1$$

   make sense – even if it is "arithmetically" not wrong.

   We *sometimes* express the applicability of these constructions by **"addability"** and **"subtractability"** of the horoi, respectively.

5. The remarkable exchange-rule has numerous applications, and it formally represents a "trick" of purposefully reworking proportions. We want to capture this property of "inheritance of similarities of proportions from one to all magnitudes" by a term of its own:

Definition 1.5 (Interchangeable Similar Magnitudes)
If it holds for two magnitudes a and $\tilde{a}$ that they have similar proportions to a third magnitude b, then this property holds with respect to all other proportions c (which form a proportion with at least one of them, a or $\tilde{a}$). We then call a and $\tilde{a}$ **interchangeable**. The following mutually equivalent **interchangeability criteria, substitution criteria,** then hold: There are equivalent

1. a and $\tilde{a}$ are interchangeable,
2. $a : b \cong \tilde{a} : b$ for any magnitude b,

(continued)

Definition 1.5 (continued)
3. $a : b \cong \tilde{a} : b$ for all magnitudes b (for which $a : b$ und $\tilde{a} : b$ exists),
4. $a : \tilde{a} \cong b : b$ for any magnitude b,
5. $a : \tilde{a} \cong 1 : 1$.

 To say, that a and $\tilde{a}$ would be "equal", as especially the condition (5) suggests, is in general neither sensible nor correct nor helpful – except in the "number-case", if a and $\tilde{a}$ are real rational or real numbers.

6. Why could we be content with the "or" in this almost casually mentioned claim within this Definition 1.5 – "which form a proportion at least with one of them, a or $\tilde{a}$"? Well, the answer is: from $\tilde{a} : b$ follows $b : \tilde{a}$, and because of $a : b$, then according to transitivity $a : \tilde{a}$ also holds. If a magnitude c is proportional to a, i.e. $a : c$ is valid, it is also proportional to $\tilde{a}$ – again because of transitivity.

Proof of Theorem 1.3

1. The inverse rule follows by applying the cross rule twice inside:

$$a : b \cong c : d \Leftrightarrow a : c \cong b : d \Leftrightarrow b : d \cong a : c \Leftrightarrow b : a \cong d : c.$$

2. We see the exchange rule like this: So if $a : b \cong \tilde{a} : b$, then according to the cross-rule $a : \tilde{a} \cong b : b$, and because for all magnitudes $c : c \cong b : b$ holds (which follows directly from the definition of similarity), again according to the cross-rule and the transitivity of similarity the statement follows:

$$a : \tilde{a} \cong b : b \cong c : c \Leftrightarrow a : \tilde{a} \cong c : c \Leftrightarrow a : c \cong \tilde{a} : c.$$

The different forms can be easily converted into each other.
3. The multiplication rules follow like so: Because $a : b \cong a : b$ holds for every proportion, adding the left side to the right side n times leads to the similarity

$$a : b \cong n * a : n * b,$$

and part (a) is proved. If we apply this to the similarity $a : c \cong n * a : n * c$, we find the rule ($b$) with the following refined application of the cross rule as well as the transitivity of the equivalence relation:

$$a : b \cong c : d \Leftrightarrow a : c \cong b : d \cong n * a : n * c \cong b : d \Leftrightarrow n * a : b \cong n * c : d.$$

So as an intermediate result we get the special case, that we can simultaneously multiply the "numerator-horoi" of a similarity-proportion. Using the inverse rule, the remainder now follows, where now the enlargement factor of the new numerator-horoi is the given number $m \in \mathbb{N}$:

$$n * a : b \cong n * c : d \Leftrightarrow b : n * a \cong d : n * c$$
$$\Leftrightarrow m * b : n * a \cong m * d : n * c$$
$$\Leftrightarrow n * a : m * b \cong n * c : m * d.$$

4. All rules concerning commensurability follow from the already existing laws, for example transitivity: So be

$$a : b \cong n : m \text{ and } b : c \cong k : j$$

commensurable proportions with natural numbers n, m, k and j. Then we can write both according to the cross and multiplication rule like this:

$$m * a : n \cong b : 1 \text{ and } k * c : j \cong b : 1.$$

So, due to the transitivity property of similarity, it follows that also

$$m * a : n \cong k * c : j \Leftrightarrow a : c \cong k * n : m * j$$

holds which means nothing else than the commensurability of $a : c$.

5. Also the piece formulas are a direct application of the three rules of Theorem 1.2, because with the cross rule

$$X : Y \cong \alpha : \beta \text{ and } X : \alpha \cong Y : \beta,$$

are equivalent and adding to the right side the left side, we get the equivalent expression

$$X : \alpha \cong (X + Y) : (\alpha + \beta),$$

which we again use the cross rule for the relation

$$X : (X + Y) \cong \alpha : (\alpha + \beta)$$

where the difference rule comes into play in the backward direction.

1.4   The Fusion of Proportions

In this section we deal with the process of merging **two proportions** to form a new **proportion.** In theoretical dealings, in computation, and in musical application, this process – **fusion** – is omnipresent. It corresponds

- in music with the **joining** ("sum", "addition") of two intervals to form a new **interval,**
- in mathematics with a **"product"** of two proportions.

To begin with a simple example: If we add to a pure major third of the proportion 4 : 5 a minor pure third of the proportion 5 : 6, and we start in reality with a fundamental c (the tonic), the third is given by e; the minor third placed on top of it leads to the note g. The note e is the final note of the major third and the initial note of the minor third, and by deleting it, we create the major interval of a fifth from c to g. In short, putting $A = 4 : 5$ and $B = 5 : 6$ together gives the result $4 : 6 \cong 2 : 3$. For this process we then use the symbolism

$$\text{fusion (product) of A and B} \equiv A \odot B = 4 : 5 \odot 5 : 6 \cong 4 : 6.$$

The **"composition"** of proportions as well of proportion chains $A \oplus B$, introduced in the next Chap. 2, on the other hand, still contains all intermediate proportions, unlike the "fusion" above: For the preceding example, it would be

$$\text{composition of A and B} \equiv A \oplus B = 4 : 5 : 6,$$

and then $A \odot B$ could be explained as the "outer" – or total – proportion of this chain.

Musically, then, the composition $A \oplus B$ stands for a "chord", namely, for example, the C major triad $c - e - g$, while the symbol $A \odot B$ stands for the total interval (here: $c - g$) achieved by the layering of both intervals: Both intervals have become one thanks to their "fusion".

The following definition now defines things in general – where we want to restrict ourselves to the case of ordinary proportions of numbers and their proportions similar to this.

Definition 1.6 (Product or Fusion of Proportions)
For two number proportions $A = a : b$ and $B = c : d$ let $A \odot B$ be the class of all proportions similar to the proportion $ac : bd$, which is given by the symbolisme

$$A \odot B \cong ac : bd.$$

We call $A \odot B$ the **"fusion of proportions A and B"** – but the expression **"product of proportions A and B"** is also common because the calculation of the fusions apparently consits of a **multiplication** of the magnitudes. But later – in chapter 5 – we will see, that this process corresponds unambiguous to **adding** exactly these intervals which are the musical representants of the related proportions. Therefore the fusion $A \odot B$ of proportions can also be regardes as to the **sum** of the corresponding intervals which we then (consequently) denote also by $A \oplus B$ – like the composition (which contains however all intermediate magnitudes of adjoining!).

More generally, let X and Y be proportions which are similar to **number** proportions A and B, and then let $A \cong X$ and $B \cong Y$. Then one defines

$$X \odot Y := A \odot B.$$

Thus fusion is also defined for all abstract proportions, which are at least similar to numerical proportions – and these are primarily the commensurable proportions.

▶   If one interprets all proportions in fraction-arithmetic form, the symbol $\odot$ is simply explained as a multiplication symbol "numerator times numerator to denominator times denominator". But note that this also includes all proportions similar to *(ac:bd)*; this corresponds to an extension of the fraction-arithmetic representation of the proportion.

In the following theorem we list the most important rules for calculus with fusions; all proofs here are immediate consequences of the definition and are based on simple fractional calculus, and we skip the details here.

Theorem 1.4 (Calculation Rules for the Fusion of Proportions)
The fusion $\odot$ is an operation, which is defined on the set of all number-proportions (or to these similar proportions) and yields in the result again a number-proportion.

1. **Invariance under similarity:** If $A_1 \cong A_2$ and $B_1 \cong B_2$
   are pairwise similar, then holds:

$$A_1 \odot B_1 \cong A_2 \odot B_2.$$

2. **Algebraic rules:** The fusion satisfies the algebraic rules of a group:
   1. The associative law applies $(A \odot B) \odot C \cong A \odot (B \odot C)$.
   2. The commutative law applies $A \odot B \cong B \odot A$.

(continued)

Theorem 1.4 (continued)

   3. The proportion $E \cong 1 : 1$ is neutral element: $A \odot E \cong E \odot A \cong A$.

   4. Each proportion A has an inverse (A^{inv}) with respect to the fusion $\odot$:

       If $A \cong a : b$, then also exactly the "inverse" proportion $A^{\text{inv}} \cong b : a$ is this

       inverse with respect to the fusion process respectively to the product.

$$A \odot A^{\text{inv}} \cong A^{\text{inv}} \odot A \cong E.$$

Corollary 1: For several number-proportions $A_1 \cong a_1 : b_1, \ldots, A_m \cong a_m : b_m$ one can determine their product – respectively their fusion – in a well-defined way:

$$A_1 \odot \cdots \odot A_m \cong (a_1 * \cdots * a_m) : (b_1 * \cdots * b_m).$$

So specifically, in the case where all proportions are equal – better: similar to a given proportion $A \cong a : b$ – the **m-fold fusion of A** is the proportion

$$A \odot \cdots \odot A = \left(\odot A \right)^m \cong a^m : b^m.$$

It corresponds to the outer (or total) proportion that would result if m given musical intervals were stacked on top of each other. This formal expression can also be usefully extended to negative exponents – written $(-m)$– if we replace the proportion A by its inverse:

$$\left(\odot A \right)^{-m} = \left(\odot A^{\text{inv}} \right)^m \cong b^m : a^m.$$

Corollary 2 (similarity proportion equations): Let A and B be given proportions. Then the – except for similarity – unique solution of the equation

$$X \odot A \cong B \, (\textit{Similarity proportion equation})$$

is given exactly by all proportions similar to the proportion $A^{\text{inv}} \odot B$. Briefly:

$$X \odot A \cong B \Leftrightarrow X \cong A^{\text{inv}} \odot B.$$

This process of fusion could certainly be conceived in a more general way, just as the chain construction discussed in the next Chap. 2 is extended beyond the case of ordinary number proportions in the case of suitable conditions. For this, one would presuppose the following two requirements as fulfilled:

Let $A = a : b$ and $C = c : d$ be two proportions, for which the magnitudes of one also builds proportions with those of the other. According to the first basic rule it is sufficient that only a magnitude of A builds a proportion with a magnitude of C – then this also follows for the others. A second requirement would be that there are similar proportions $\widetilde{A} = \widetilde{a} : \widetilde{b}$ and $\widetilde{C} = \widetilde{c} : \widetilde{d}$ to both A and C, so that their connecting magnitudes ($\widetilde{b}$ and $\widetilde{c}$) are in a $1 : 1$ – proportion, i.e.: $\widetilde{b} : \widetilde{c} \cong 1 : 1$. Then the proportion $\widetilde{a} : \widetilde{d}$ could be defined as the product of A and C, and then $\widetilde{a} : \widetilde{d}$ could be interpreted as the "fusion" of the proportions $\widetilde{A}$ and $\widetilde{C}$ which are similar to A and C.

But one suspects, that in this generality already the elementary calculating-rules of the preceding theorem, mentioned for the product, would be a more complicated matter. For practical use it is sufficient to have the fusion for number-proportions at disposal, and by the process of similarity already far reaching abstract extensions are possible and applicable without any additional considerations in a similar uncomplicated way as it is the case for numbers.

We now add some examples of musical constructions:

Example 1.2 Fusions for Iterated Interval Layering

According to the rule of the preceding theorem for layering with a basic interval – or with a given proportion – the musical intervals result:"

1. (Fifth $2 : 3$) $\odot$ (Fifth $2 : 3$) $\cong$ major Ninth $4 : 9$.
2. If this ninth is shortened by an octave, the Pythagorean whole tone "*Tonos*" is created to the proportion $8 : 9$, for this is the fusion (Ninth $4 : 9$) $\odot$ (Octave $1 : 2)^{inv} = (4 : 9) \odot (2 : 1) = $ Tonos $8 : 9$.
3. (Tonos $8 : 9$) $\odot$ (Tonos $8 : 9$) $\cong$ Phythagorean third"Ditonos"$64 : 81$.
4. Tonos $\odot$ Tonos $\odot$ Tonos $= (\odot \text{Tonos})^3 = $ "Tritonos" $\cong 2^9 : 3^6$.

   The *tritonos* (or tritone) thus consists of three (here: Pythagorean) whole tone steps. This interval can be used very conveniently to calculate the leading semitone of the Pythagorean octave scale – the *limma*: If we imagine the difference of tritone to fifth, the tone model $c - f\# - g$ is less suitable; rather, we should think "Gregorian": on white keys, the total proportion of the span

   $$f - g - a - h \text{ respectively fa } - \text{ sol } - \text{ la } - \text{ si(Gregorian note)}$$

   in Pythagorean fifth tuning is exactly the tritone formed above.

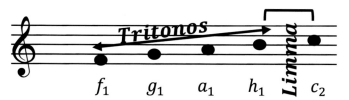

We determine the difference of h to its following tone c' : If we denote this semitone interval with the symbolic unknown X, then – because $f - c'$ forms a fifth – the similarity equation applies

$$X \odot \text{Tritonos} \cong \text{Fifth} \Leftrightarrow X \cong (2 : 3) \odot (3^6 : 2^9) = 3^5 : 2^8 = 243 : 256.$$

Thus, using this balance, we have found the proportion of the semitone "*Limma (L)*" by means of the similarity equation.

The Limma $243 : 256$ divides the tonos "only to about 45%" – if one uses the metric scale of the cent measure. Its partner in the tonic, the *Apotome (A)*, thus makes up the (audibly) larger remainder. Certainly the octave balance

$$2 \text{ Limma } + 5 \text{ Tonos} = \text{Octave } - \text{better:}$$
$$2 \text{ Limma } \oplus 5 \text{ Tonos} = \text{heptatonic Octave scale}$$

would also lead to the same result – but this is a "quadratic" equation for the limma, because a doubling according to our Corollary 1 of the theorem leads to a squaring. Of course, one still quickly finds the integer solution here as well

$$L \cong 243 : 256,$$

which is recommended as a practicing task.

5. ($\odot$ Ditonos $64 : 81)^3 \cong 2^{18} : 3^{12} = 262.144 : 531.441,$

and this last value, which is obviously numerically a little "smaller" than the proportion $1 : 2$ (which is why this interval is larger than an octave(!)), deviates from the octave by the famous **Pythagorean comma** – in our current language by the proportion $524.288 : 531.441$. Because now we can ask by means of the similarity proportion equation: For which proportion X *is* the balance valid:

$$(1 : 2) \odot X \cong 262.144 : 531.441?$$

Then the solution of this similarity proportion equation is:

$$X \cong (2 : 1) \odot (262.144 : 531.441) \cong 524.288 : 531.441 = 2^{19} : 3^{12}.$$

In short, three Pythagorean major thirds (or six Pythagorean whole tones) exceed the octave by the small interval called the "Pythagorean comma," which has the proportion 524.288 : 531.441 and is about the size of a quarter of an ordinary semitone step.

◀

Fig. 1.5 Sketch of the symmetry principle of the geometric centre

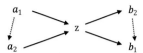

Note We will encounter more of this calculation with proportions in the last Chap. 5; there, the systematic nature of this arithmetic of proportions leads us to a form of calculation with musical intervals and their scales and balances that is both convenient and conceptually clear.

With the help of this approach, however, we can also prove, in consistent application of the laws of proportion, a principle of symmetry which – as we shall see in Sect. 3.5 – establishes the characteristic property of the geometric mean as the centre of symmetry of a chain of proportions. Figure 1.5 may illustrate the situation:

Theorem 1.5 (Symmetry Principle of the Geometric Center)
Suppose we have magnitude pairs (a_1, b_1) and (a_2, b_2) and a magnitude z, which – as sketched in Fig. 1.5 – satisfy the two symmetries of proportions:

1. $a_1 : z \cong z : b_1$
2. $a_2 : z \cong z : b_2$.

For the sake of simplicity, we assume that all proportions are numerical proportions or similar to them.

Then also the crosswise ordered magnitude pairs themselves are in a similarity proportion to each other, because the similarity equation is valid

$$a_1 : a_2 \cong b_2 : b_1.$$

Conclusion: If a magnitude z is the geometric center for n pairs of magnitudes (a_k, b_k), $(k = 1, \ldots, n)$, which means that

$$a_k : z \cong z : b_k \text{ for all indices } k = 1, \ldots, n$$

all pairs of magnitudes are related to each other by the proportions

$$a_k : a_m \cong b_m : b_k \text{ for all index pairs } k, m \in \{1, \ldots, n\}$$

in a reciprocal (and crosswise) relationship.

(continued)

Theorem 1.5 (continued)

We say that z is the **geometric mean (geometric center)** of a_1 and b_1 *as well as of* a_2 *and* b_2 and possibly of all other data pairs. And more generally, we will say later that z *is* the **center of the proportion chain**

$$a_1 : a_2 : \ldots : a_n : b_n : \ldots : b_2 : b_1$$

and that this proportion chain is symmetrical and thus similar to its "reciprocal".

Proof After defining the fusion, we have the following equations using the shortening rule:

$$(z : a_1) \odot (a_1 : a_2) \cong z : a_2 \Leftrightarrow (a_1 : a_2) \cong (z : a_2) \odot (z : a_1)^{\text{inv}} \cong (z : a_2) \odot (a_1 : z),$$
$$(z : b_2) \odot (b_2 : b_1) \cong z : b_1 \Leftrightarrow (b_2 : b_1) \cong (z : b_1) \odot (z : b_2)^{\text{inv}} \cong (z : b_1) \odot (b_2 : z).$$

If we now set the preconditions, the lower of the two equations is valid, for example

$$(z : b_1) \odot (b_2 : z) \cong (a_1 : z) \odot (z : a_2),$$

because it is then $z : b_1 \cong a_1 : z$ and with the inversion rule also $b_2 : z \cong z : a_2$. However, we can replace all proportions by similar ones in fusions, and the result remains unaffected. But from both equations it follows that then also the proportions $a_1 : a_2$ and $b_2 : b_1$ are similar, and the proof is complete.

The corollary is apparently a constant duplication of these relations to two arbitrarily chosen pairs of magnitudes (a_k, b_k) and (a_m, b_m), with $k, m \in \{1, \ldots, n\}$.

Proportion Chains

2

> *only in this way can it be understood that without harmony no science can be complete (perfecta), because without it there is nothing.... (Aristides Quintillianus, from Flotzinger [5], p. 128).*

Now we will decisively expand the concept of proportions to be able to use it to describe the most important applications with regard to ancient music theory.

▶   If we join one proportion to another – in such a way that the back member of the former is identical with the front member of the latter – then we get what is called a **Chain of Proportions.**

One encounters these chains of proportions both in the ancient centuries, where they have methodically appropriated the sprawling field of internal mathematical relations, and in music-theory. Here, there are three significant areas in particular:

- **Chordal-theory:** After all this is the layering of intervals – literally, the joining of proportions – while retaining all magnitudes.
- **The scale-theory:** The stringing together of – as a rule – commensurable steps to form a scale inherently accommodates the complete set of instruments of the proportion chains and profits in many ways from their ordering laws.
- **Modology:** The study of keys and their inner musical functional structures especially the variety of ancient tonalities, ecclesiastical keys as well as playful variants and new creations – can also make ample use of the structural laws of the theory of proportions and thus reveal interesting connections between them.

© Springer-Verlag GmbH Germany, part of Springer Nature 2024
K. Schüffler, *Proportions and Their Music*,
https://doi.org/10.1007/978-3-662-65336-4_2

The text of this chapter on proportion-chains is structured by the following sequence, in which we embed and develop the mathematical theory of multilevel iterations of proportions:

1. The basic terms of proportion chains,
2. the composition of proportion chains,
3. the reciprocal proportion chains,
4. the symmetrical proportion chains,
5. the proportion chain theorem.

This last section describes the properties of similarities, inversion, adjoint (joining) and mutual interplay in a mathematically very abstract way – but in doing so we achieve the goal of greatest possible generality and we gain useful structures and architectures for symmetrical chains of proportions.

> The reward for these efforts is also ultimately – as is more often the case – the gain of more comprehensive connections, without any detailed and therefore also usually laborious calculations.

2.1   Basic Concepts for Proportion Chains

The goal of this first section is to explain the most important concepts surrounding the topic of multilevel proportion sequences – which we will refer to as proportion-chains throughout. We start with the simplest situation of the unmediated **J**uxtaposition of two proportions to form a chain of proportions:

▶   **Important**
   If *a, b, c* are magnitudes ('magnitudes, horoi'), so that the two proportions $a : b$ as well as $b : c$ exist, one writes for this the concise expression $a : b : c$, for short

$$a : b : c \Leftrightarrow a : b \text{ and } b : c$$

'*a* is in proportion to *b* and *b* is in proportion to *c*'. Therefore, due to the transitivity of the proportion property, there is also the proportion $a : c$. One calls then $A = a : b : c$ a **2-step** – or also **3-member – Proportion Chain.**

This concept now finds a general form, and we follow up with a first detailed definition, which aims to explain to us the vocabulary of multilevel proportion constructs:

Definition 2.1 (Proportion-Chains and Their Basic Concepts)
An **n-step proportion chain** is defined by the statement

$$A = a_0 : a_2 : \ldots : a_{n-1} : a_n \Leftrightarrow a_0 : a_1 \text{ and } a_1 : a_2 \text{ and } \ldots \text{ and } a_{n-1} : a_n$$

fixed. Thus, an n-step proportion chain $A = a_0 : a_1 : \ldots : a_n$ consists of the ordered sequence of n proportions $a_k : a_{k+1}$ ($k = 0, \ldots, n-1$), which we will also refer to as the **'step proportions'** of **the proportion chain A in** this context. Alternatively, one also encounters the expression of an **($n + 1$)-membered proportion chain**; **1-step proportion-chains** are ordinary proportions $a : b$.

1. Due to the reflexivity, symmetry and transitivity for proportions, it is true that all magnitudes of a proportion chain $A = a_0 : a_1 : \ldots : a_n$ form the proportions $a_k : a_j$ $(k, j = 0, \ldots, n)$ among themselves.
2. If the order of the proportions elements is reversed, the **inverse proportion chain** to A **is** obtained

$$A^{\text{inv}} = a_n : a_{n-1} : \ldots : a_1 : a_0,$$

   whose step proportions are the inverse step proportions of A with a simultaneous inversion of their order.
3. In an n-step proportion chain $A = a_0 : a_1 : \ldots : a_n$ is called
   1. the magnitude a_0 **Front Proportional** or **Initial Member**,
   2. the magnitude a_n **Posterior Proportional** or **End Member**,
   3. both magnitudes a_0 and a_n **outer magnitudes** and their proportion $a_0 : a_n$ the **Total Proportion** of A,
   4. all other magnitudes in between **Mean Proportional, Intermediate,**
   5. the magnitude pairs a_k and a_{n-k} situated 'in opposition' with $0 \leq k \leq n$, thus

   $$(a_0 \text{ and } a_n),\ (a_1 \text{ and } a_{n-1}),\ (a_2 \text{ and } a_{n-2}),\ \ldots, (a_n \text{ and } a_0),$$

   magnitudes **Mirrored to each other** – or also magnitudes in **Diametrical** or **Contra-Position**. In this context we then set the star symbol and write

   $$a_k^* = a_{n-k} \quad (\text{for } 0 \leq k \leq n).$$

4. A k-level proportion chain B is called an (ordered) **Sub- or Partial Proportion chain of A**, if its members are a partial selection of the members of A and if

(continued)

Definition 2.1 (continued)

the order (sequence) of the chain A is observed. Often, we use the simple word "Subchain" for this. Mathematically, this is expressed by the formalism

$$B \subseteq A \Leftrightarrow B = a_{n_1} : a_{n_2} : \ldots : a_{n_k}$$

$$\text{with } 0 \le n_1 < n_2 < \ldots < n_k \le n$$

and the subset symbol' $\subseteq$'.

5. The uniquely existing chain for such a partial proportion chain B, consisting of its mirrored magnitudes and adhering to the magnitude order defined by the upper chain A, is the partial proportion-chain B^*, which is also signed by a star symbol and is defined as follows

$$B = a_{n_1} : a_{n_2} : \ldots : a_{n_k} \Leftrightarrow B^* = a_{n_k}^* : \ldots : a_{n_1}^*.$$

It is called the **(Diametrical) Subchain mirrored to** the **Subproportion-chain B.**

6. A proportion chain is called **Commensurable** if all its step proportions are commensurable. Consequently, all further proportions $a_k : a_j$ of the chain are also commensurable due to the transitivity property.

7. A proportion chain is called a **Rational** (respectively **real**) **number proportion chain** $\Leftrightarrow$ all quantities a_k are themselves already rational (respectively real) numbers.

8. In cases like number-proportions, when magnitudes a_0, a_1, ..., a_n can be ordered, we can add the following terms: The proportion chain $A = a_0 : a_1 : \ldots : a_{n-1} : a_n$ is called **Ascending-ordered** if

$$a_0 < a_1 < \cdots < a_n$$

is valid. The case of equality is traditionally excluded here, but it would be an unproblematic generalization. Analogously, one defines **'descending-ordered'**.

Some Remarks

1. As mentioned, the literature also knows the notion of **m-membered proportion chains**. For example the two-level chain $a : b : c$ is a 3-member chain and in general

$A = a_0 : a_2 : \ldots : a_{n-1} : a_n$ is an $(n + 1)$-member chain. As a result the simple relation says that a **n-level proportion chain** is an **$(n + 1)$-member proportion chain.**

The background of the use of these differing terms is essentially: In one case, the step case, one sees the chain as composed of a number (n) of 'proportions', in the other case one has rather in mind the number $(n + 1)$ of 'magnitudes', which are also called "members" of the chain.

In our – musically motivated – considerations, the proportions themselves play a more decisive role (are key), less the values of the magnitudes. From a mathematical point of view, too, the 'step' designation proves to be advantageous and optimally and more conclusively suited for a closed, coherent representation.

Note: Therefore we also preferred the notation $A = a_0 : a_1 : \ldots : a_n$ to a notation $A = a_1 : a_2 : \ldots : a_{n+1}$, which also has advantages in the context of including mirrored magnitudes: Because $a_k^* = a_{n-k}$ (with $k = 0, \ldots, n$)   reads   better   than $a_k^* = a_{n+2-k}$ (with $k = 1, \ldots, n+1$).

2. It is obviously important for the formation of a chain that the rear member of a proportion is the front member of another proportion at the same time/simultaneously, so that a sequence of successive step proportions is (finally) formed.

   We will see in the next Sect. 2.3, that exactly this condition – in its generalized form – is the decisive one, which we need for generating proportion chains or chains from other partial chains by means of an composition procedure. This generalized form consists in the fact that it is sufficient for this purpose (weakening) to require, the posterior member of the anterior proportion (or chain) forms with the anterior member of the posterior proportion (or chain) a proportion to be of the form $1 : 1$ (if the two are not identical anyway), that is, that the two members must be interchangeable. By using the exchange rule, the chain construction can succeed, but we will go into this later when we present the chain construction method in detail.

3. The concept of diametrical or mirrored pairs of magnitudes may seem a little daunting in the general mathematical formulation of indexing: However it is the simple process of exchanging the counting of magnitudes from the front with that from the back: The first magnitude corresponds to the last, the second to the second last, the third to the third last, and so on. In other words:

   If we write the two proportion chains A and A^{inv} on top of each other then the magnitudes immediately above each other are diametrical or mirrored.

   In musical applications, mirrored magnitudes – or, more generally, mirrored-chains of partial proportions – play an important role. We can connect major and minor as well as many other things of the scale architectures with them. And this is precisely the reason why we include these terms.

4. For the constructive elements above the architecture of proportion chains there are some rules, which can be read from the definitions without (much) effort:

   (a) For any proportion chain A, $A \subseteq A$ is also a subchain of itself.

   (b) For each magnitude x is $(x^*)^* = x$, for each subchain $B \subseteq A$ is $(B^*)^* = B$.

(c) For the construction of a mirrored chain B^* the following applies

Rule: If a magnitude x belongs to the subchain B, there are two cases:
The mirrored magnitude x^* also belongs to B – then both magnitudes belong
to B^*. Or x^* does not belong to B – then x^* belongs to B^*, but not x.

5. The **Proportion-measure of** a number-proportion-chain $A = a_0 : a_1 : \ldots : a_n$ is – as we
herewith determine it – itself again a **proportion** – namely its **total proportion.** As a
result this measure is obviously the fusion of all step-proportions $(a_0 : a_1)$, …, $(a_{n-1} : a_n)$, in formulas

$$a_0 : a_n = (a_0 : a_1) \odot \ldots \odot (a_{n-1} : a_n),$$

which in turn is a consistent application of the merger laws.

▶ **Important**

In fact, as mentioned several times, pure musical intervals are usually expressed in
their proportion measure – such as 1 : 2 for the octave, 2 : 3 for the fifth, and so on –
rather than as mere numerical values. The proportion conveys (in this case) much
more than just a numerical value – such as 0.66... for a pure or pythagorean fifth.
Incidentally, this indication would not have been historically formulated at all.

On the other hand, the **frequency measure of** the 'interval' $[a, b]$ is the quotient
b/a – i.e. the ratio of **output to input** – and then the **frequency measure of the
number-proportion-chain A** is the product of the frequency measures of its step
intervals, like the fractional arithmetic

$$\frac{a_n}{a_0} = \frac{a_n}{a_{n-1}} * \frac{a_{n-1}}{a_{n-2}} * \cdots * \frac{a_1}{a_0}$$

and the result is the frequency measure of the total proportion. This formula is already a
special case of the general multiplicative property of proportion and frequency measure at
the same time.

Example 2.1 Proportion-Chains

Given is the concrete proportion chain

$$A = 1 : 2 : 3 : 4 : 5 : 6,$$

which we already have become acquainted with the chain designated in ancient times by
the term **'Senarius'** – see the section 'For Use' at the beginning of the book. This
six-membered chain has the five step proportions

$$1 : 2 \text{ and } 2 : 3 \text{ and } 3 : 4 \text{ and } 4 : 5 \text{ and } 5 : 6.$$

Initial member is magnitude 1 and final member is 6; the total proportion is the ratio $1 : 6$ and corresponds to the musical interval of a pure fifth $2 : 3$ enlarged by two octaves $1 : 2$. The chain inverse to A, A^{inv}, is the backward proportion arrangement

$$A^{inv} = 6 : 5 : 4 : 3 : 2 : 1.$$

For example, a two-level subchain of A would be $B = 2 : 4 : 5$; its mirrored magnitudes are 5, 3, and 2, so we get the mirrored chain $B^* = 2 : 3 : 5$.

Another example would be $B = 1 : 2 : 3 : 5$, and then $B^* = 2 : 4 : 5 : 6$ would be. ◀

Many of the terms profound to music-theory have very specific chains of proportions as their substantive points of reference paired with their accompanying vocabulary. The constructions 'tetrachord' and 'scale' serves as a good example. In the language of proportion chains, this now reads as follows:

Example 2.2 Proportion Chains and Musical-Scales

A **musical-scale** is a sequence of tones ordered in ascending order (by pitch); for $(n + 1)$ tones, therefore, there is an n-step sequence of intervals (the 'step intervals') of the scale, and we get the proportion chain

$$A = a_0 : a_1 : \cdots : a_n,$$

for which the total proportion $a_0 : a_n$ is the circumference or also the **Ambitus** of the scale. Whether the magnitudes a_k represent pitches (in hertz or the like) or whether the steps are expressed by numerical ratios is of no importance.

Often scales are '**Octave-scales**'; the range is one octave $1 : 2$. But other scales are also inherent to music-theory in terms of structure; well-known examples would be:

1. A **Tetrachord** is an (ascendingly ordered) 3-step numerical proportion-chain of the extent of a pure fourth – i.e. a proportion-chain $A = a_0 : a_1 : a_2 : a_3$, of 4 members (tones) for which the total proportion is $a_0 : a_3 \cong 3 : 4$ and thus corresponds to a perfect fourth (the musical complement of the fifth in the octave).
2. The scale is called **Pentatonic** ⇔ it has 4 steps (so 5 members, tones), and the total range is (usually) a major sixth.

3. The scale is called **Heptatonic** ⇔ it is an octave scale of 8 tones and therefore has **7** step proportions.

4. The scale is called **Chromatic** ⇔ it is a 12 – step octave scale and therefore has **13** members, tones as magnitudes. The octave tonic (fundamental) is therefore added.   ◄

These examples alone serve to show that our concept of proportion chains – and the corresponding conceptual environment – forms a backbone of traditional music- theory.

Similarity of Proportion Chains

In almost all calculations and construction processes involving, these are inevitably changed; either numerical values of individual proportions can change, or the step arrangements can vary. In this context it is important to realize, that all often necessary changes are at least meaningless for applications, if the changed structures are 'similar' to the initial ones. This is how it works in geometry: In similar triangles, the same relationships and conditions take place. The notion of similarity has already been introduced for ordinary proportions in the preceding Chap. 1 (see Definition 1.3) and now we transfer it to serial arrangements of proportions. In doing so we can use three criteria that are equivalent to each other, which can be used as a guide as needed.

Definition and Proposition 2.2 (Similarity of Proportion Chains)

Two n-step proportion chains

$$A = a_0 : a_1 : \ldots : a_n \text{ and } A' = a'_0 : a'_1 : \ldots : a'_n$$

are called **similar** – and one writes then

$$a_0 : a_1 : \ldots : a_n \cong a'_0 : a'_1 : \ldots : a'_n - \text{respectively} : A \cong A' -,$$

if one of the following criteria is verifiable:

1. **Step criterion:** All equally positioned step proportions of both chains are similar; thus, for all $k = 0, \ldots, n - 1$ the similarities are valid.

$$a_k : a_{k+1} \cong a'_k : a'_{k+1} \text{ viz.} : \quad a_0 : a_1 \cong a'_0 : a'_1, \ldots, a_{n-1} : a_n \cong a'_{n-1} : a'_n.$$

2. **Magnitude criterion:** All ordered – i.e. corresponding in the enumeration – magnitude proportions are similar.

(continued)

Definition and Proposition 2.2 (continued)

$$a_0 : a_0' \cong a_1 : a_1' \cong \cdots \cong a_n : a_n'.$$

3. **Value criterion:** For real (number) proportion chains, the following holds: There is a positive factor λ such that for all k the simultaneous equation: $a_k' = \lambda a_k$, holds – briefly:

$$\left(a_0', a_1', \ldots, a_n'\right) = \lambda(a_0, a_1, \ldots, a_n) \Leftrightarrow A' = \lambda A.$$

Proposition (mathematics of similarity of proportion chains):
1. These three criteria are equivalent to each other, that is: Step criterion $\Leftrightarrow$ Magnitude criterion $\Leftrightarrow$ Value criterion.
2. For all n-level proportion chains A, B, and C, the properties hold:
   (a) **Reflexivity:** Each chain is similar to itself: $A \cong A$.
   (b) **Symmetry:** If B is similar to A, then A is also similar to B: $A \cong B \Leftrightarrow B \cong A$.
   (c) **Transitivity:** If A is similar to B, B is similar to C, then A is also similar to C:

$$A \cong B \text{ and } B \cong C \Rightarrow A \cong C.$$

Thus, the similarity has the characteristics of an **Equivalence Relation** on the set of all n-level proportion chains.

Some Remarks

1. It is not difficult to see that these three forms of description are equivalent (of course). Again, the cross-rule is responsible for this. Likewise, it is clear that similarity requires the same number of steps. Proportion chains with different numbers of steps cannot be inherently similar.
2. In modern language the statement of the Proposition means: Because the similarity on the set of all (n-level) proportion chains causes an **equivalence relation,** consequently a division of this set into disjoint (=nonintersecting) classes arises.

Thus, a class consists precisely of all the chains of proportions that are similar to one of its representatives – and thus to all other of its members, in a symbolical form:

▶ **Important**

  If $A = a_0 : a_1 : \ldots : a_n$ is an n-step proportion chain, then

$$\mathcal{M}(A) = \{A' = a_0' : a_1' : \cdots : a_n' \mid a_0' : \cdots : a_n' \cong a_0 : \cdots : a_n\}$$

is the set of all proportion-chains A' that are similar to the proportion-chain A. As an result all proportion-chains contained in $\mathcal{M}(A)$ are similar to each other due to the transitivity of the property 'similar'. Accordingly, non-similar proportion chains A and B also have non-overlapping classes $\mathcal{M}(A)$ and $\mathcal{M}(B)$.

In the proportion-chain theorem of Sect. 2.5 we want to go into the structure of the division into equivalence classes again and prove it.

Although the mathematics described in the preceding remark (2) seems to have only *a remote 'practical'* significance and can certainly give new progress to the reports of a so called 'grey theory'. Nevertheless, it is firmly implanted in everyday practical use, As an example we enter the purely diatonic major chord in the tonal position with

$$20 : 25 : 30.$$

This is **musically the same** as the proportions

$$4 : 5 : 6 \ \text{ or } \ 8 : 10 : 12 \ \text{ or} \ldots \text{or} \ldots,$$

as we already find it at the extended Pythagorean canon $6 - 8 - 9 - 10 - 12$, the diatonic canon. Thus the diatonic major triad – beginning on the tonic – is basically the object of the complete equivalence class $\mathcal{M}(4 : 5 : 6)$.

Two more brief explanatory examples follow:

Example 2.3 Similarities for Proportion-Chains

1. **The Major Chord:** The two chains $A = 4 : 5 : 6 : 8$ and $B = 20 : 25 : 30 : 40$ are similar, as all three criteria show.
2. **Geometry:** If F_1 is the area of the incircle (with radius r) of the equilateral triangle Δ ABC (with the side length a), F_2 is the area of the triangle Δ ABC, F_3 is the area of the outer rectangle and F_4 is the area of the circumcircle (with radius R), then firstly we have the proportion $r : a/2 = 1 : \sqrt{3}$ and the equality $R = 2r$ (the latter is true because of the special geometry of the equilateral triangle) and then we become the proportion chain

Fig. 2.1 Inside and outside
radii in an equilateral triangle

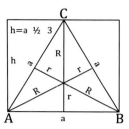

$$F_1 : F_2 : F_3 : F_4 \cong \frac{1}{12}\pi a^2 : \frac{1}{4}\sqrt{3}a^2 : \frac{1}{2}\sqrt{3}a^2 : \frac{1}{3}\pi a^2$$

$$= a^2\left(\frac{1}{12}\pi : \frac{1}{4}\sqrt{3} : \frac{1}{2}\sqrt{3} : \frac{1}{3}\pi\right) \cong \frac{1}{12}\pi : \frac{1}{4}\sqrt{3} : \frac{1}{2}\sqrt{3} : \frac{1}{3}\pi.$$

Figure 2.1 illustrates these formula relationships using the Pythagorean theorem. This chain
of proportions is, moreover, ascending, as shown by the rounded numerical values:

$$F_1 : F_2 : F_3 : F_4 \approx 0,2618 : 0,4330 : 0,8660 : 1,0472.$$

◀

The sketch in Fig. 2.1 may serve to illustrate the geometric relationships of the inner and
outer circles in an equilateral triangle.

2.2   The Composition of Proportion Chains

Thus, by defining, proportion chains are created by joining together individual proportions
or, more generally, two given chains of proportions, insofar as this is possible. Here an
example from music offers itself:

Let us suppose we had the two proportions 3 : 5 and 4 : 7. The first proportion stands –
musically – for a major sixth, the second for a pure seventh (natural seventh). Assuming
that we would layer these two intervals on top of each other – first the sixth, then the
seventh – the musical result would be a triad. But how can its assembled proportions be
represented in a chain? Well, we expand (multiply) the first proportion by 4, the second by
5, and from the two new proportions 12 : 20 and 20 : 35 – which are similar to the old
ones – we then create the two-step proportion-chain 12 : 20 : 35, which also simultaneously
provides the proportions of the front and back members 12 : 35. This construction was
therefore based on it.

▶   We brought the final magnitude of the first and the initial member of the second
      proportion to the same size – abstractly: To the proportion 1 : 1 – by a respective
      integer multiplication of both proportions, and then simply by merging the now equal
      middle magnitude constructed a 2-step chain whose proportions are similar and in
      order to the given ones.

Wherever we layer intervals or chords on top of each other, proportions or already
existing chains are brought together. This joining together corresponds to the musical
construction of the **adjunct** (the 'layering on top of each other') of intervals – while
retaining the 'intermediate notes'. We also refer to this as **composition of proportions** – or
chain composition – and consequently use the same formula symbol ('⊕') for both. The
following general conceptualizations now serve for this purpose:

**Definition 2.3 (The Composition - as a form of Adjunction ($A \oplus B$)
of Proportion Chains)**
Given two n- and m-level proportion chains A and B, respectively,

$$A = a_0 : a_1 : \cdots : a_n \text{ and } B = b_0 : b_1 : \cdots : b_m,$$

with the respective n and m step proportions $a_k : a_{k+1}$ and $b_j : b_{j+1}$.
      Then B is called **connectable to A** if it has an $(n + m)$-step proportion chain C

$$C = c_0 : c_1 : \ldots : c_n : \ldots : c_{n+m-1} : c_{n+m}$$

whose levels have the $(n + m)$ proportions

$$c_0 : c_1 \cong a_0 : a_1, \ldots, c_{n-1} : c_n \cong a_{n-1} : a_n$$
$$c_n : c_{n+1} \cong b_0 : b_1, \ldots, c_{n+m-1} : c_{n+m} \cong b_{m-1} : b_m$$

satisfy. Every such $(n + m)$-step-proportion-chain is then called **an Composition -
sometimes also an Adjunction or Composition - of B to A**. All $(n + m)$ step-
proportions of C are thus similar to all step-proportions of A and B, and their ordering
proceeds in this order:

      The front n–level subchain $c_0 : c_1 : \cdots : c_n$ *of* C is similar to the chain A,

      the trailing m–step subchain $c_n : c_{n+1} : \cdots : c_{n+m}$ *of* C is similar to the chain B;

and in the common magnitude c_n both chains A and B – i.e. similar copies of each of
them – are joined together.

(continued)

> **Definition 2.3** (continued)
>
> From this we immediately recognize two invariants under similarity: if C is a composition of B at A, then this holds for any chain C' with $C' \cong C$. We then use the symbol $A \oplus B$ for the set of all such compositions, symbolically
>
> $$A \oplus B = \{C \mid C \text{ is a } (n + m) - \text{stepped chain with the above proportions}\},$$
>
> so that a concrete composition C is an 'element of this set'.

Nevertheless, we agree to use the catchier form $C \cong A \oplus B$ for a single chain, which is an composition of B to A, instead of using the admittedly consistent set-theoretic notation '$C \in A \oplus B$'.

Altogether as an result the expression $C \cong A \oplus B$ means that the chain of proportions B can be attached to the chain A. As an example the chain C is an element of $A \oplus B$, also called a concrete composition.

For the important special case of two simple proportions $A = a : b$ and $B = c : d$, which are, after all, the shortest possible chains of proportions, would be with two similar proportions connectable using a common magnitude,

$$x : y \cong a : b \text{ and } y : z \cong c : d,$$

a composition $C = A \oplus B$ is given as a 2-step proportion chain $x : y : z$ with the 2 step proportions similar to A and B, respectively.

In the following we will see this specific case for pure 'number proportion chains', i.e. those that we encounter almost exclusively in music-theory,. The process of joining two proportion-chains is basically very simple. For the general case of 'abstract' proportion-chains, on the other hand, the situation is unfortunately somewhat more complicated. This is the case because, in an composition, the initial member (b_0) of the back chain (B) coincides with the final member (a_n) of the front chain (A); they are 'identified'. In the case of numbers, this is (pretty) easy: by multiplying the magnitudes of one of the two chains, we can always establish equality and the composition chain is complete, but in the general case the necessary identification of these two 'connecting-magnitudes' $(a_n$ and $b_0)$ is done first by bridging similar chains, in order to be able to use the exchange-procedure of Theorem 1.3 afterwards. We also illuminate this more abstract subject and for this we are helped by some further deepening notions about the connectivity of proportions, which are defined as follows:

Definition 2.4 (Commensurable Connections of Proportion Chains)
Given are an n- and an m-level proportion chain

$$A = a_0 : a_1 : \ldots : a_n \text{ and } B = b_0 : b_1 : \ldots : b_m.$$

1. The chain B is called **'Directly connectable with the chain A'** if there is an 'equality proportion' for the **Connection Magnitudes** a_n and b_0, i.e. if

$$a_n : b_0 \cong 1 : 1$$

   holds, so that according to Theorem 1.3 these connection magnitudes are also interchangeable.
2. The chain B is called **'Indirectly-connectable with the chain A'**, if there are at least similar proportion chains A' and B' with $A' \cong A$ and $B' \cong B$, so that the chain B' is directly-connectable with the chain A': There is the equality proportion for their connecting magnitudes,

$$a'_n : b'_0 \cong 1 : 1.$$

3. The chain B is called **'commensurably direct-connectable'** with the chain A, if there is a commensurable proportion for the connection magnitudes a_n and b_0, that is:

$$a_n : b_0 \cong p : q \text{ with natural numbers } p \text{ and } q.$$

4. Finally, the chain B is called **'commensurably indirectly Connectable'** with **the chain A,** if there are at least similar proportion chains A' and B' with $A' \cong A$ and $B' \cong B$, so that the chain B' is commensurably directly connectable with the chain A'; thus, there is a commensurable proportion for their connecting magnitudes.

$$a'_n : b'_0 \cong p : q.$$

Some Remarks

1. We show later that it is exactly the connectivity condition (2.), which is necessary and sufficient for the existence of the whole family of all similar composition-chains of two proportion-chains.
2. These four concepts are partially implicit among themselves; we can easily see the implications of the criteria

$$(1.) \Rightarrow (2.) \text{ and } (1.) \Rightarrow (3.) \text{ and } (1.) \Rightarrow (4.) \text{ and } (2.) \Rightarrow (4.) \text{ and } (3.) \Rightarrow (4.).$$

Therefore, the criterion (4.) is the most general form of these connecting terms. It is very interesting to see that we can show the equivalence of (4.) and (2.): This is done in Theorem 2.1 below.

3. Another consequence of these connectivity notions is the fact – especially significant in musicology – that any commensurable n-level proportion chains can be treated as n-level integer chains, since they are namely similar to those. We also show this in the following Theorem 2.1.
4. Another important application of these general notions of connection is the fact that two commensurable proportion chains A and B are always indirectly connectable – both A with B and B with A. For by definition commensurable proportion chains are similar to number proportion-chains and for these the merging process is possible.

Theorem 2.1 (Composition Criteria for Proportion Chains)
The following criterion provides information about the possibilities of composing two chains $A = a_0 : a_1 : \ldots : a_n$ and $B = b_0 : b_1 : \ldots : b_m$ to a chain $C \cong A \oplus B$:

1. The composition criterion
   They are equivalent:
   (a) There is a composition $C \cong A \oplus B$ ('B is attachable to A').
   (b) The chain B is commensurably indirect-connectable with the chain A.
   (c) The chain B is indirectly connectable with chain A.
   Accordingly, there are then proportion chains A' and B' with $A' \cong A$ and $B' \cong B$,

$$A' = a'_0 : a'_1 : \ldots : a'_n \text{ and } B' = b'_0 : b'_1 : \ldots : b'_m,$$

so that for their connection magnitudes the 'Composition condition'

$$a'_n : b'_0 \cong 1 : 1$$

is fulfilled. In particular, the properties (2) and (4) of definition (2.4) are equivalent.

(continued)

Theorem 2.1 (continued)

2. Composition formulas for proportion chains

    (a) Each n commensurable proportions can be combined into an n-level (commensurable) chain of proportions.

    (b) Corollary: Every commensurable n-step proportion chain is similar to an integer n-step proportion chain, symbolically:

$$A = a_0 : a_1 : \cdots : a_n \text{ is a commensurable proportion chain} \Leftrightarrow$$

$$A \cong m_0 : m_1 : \cdots : m_n \text{ with natural numbers } m_0, m_1, \ldots, m_n.$$

    (c) The **first composition formula**: Let there be n commensurable number proportions

$$A_1 = a_1 : b_1, \ A_2 = a_2 : b_2, \ \ldots, A_n = a_n : b_n$$

    with real numbers a_k, b_k $(k = 1, \ldots, n)$ are given. Then the n-step proportion chain C, defined by

$$C \cong A_1'' : A_2'' : \cdots : A_n'' \equiv (a_1 * a_2 * \cdots * a_n) : (b_1 * a_2 * \cdots * a_n) :$$
$$(b_1 * b_2 * a_3 \cdots * a_n) : \cdots : (b_1 * b_2 * \cdots * b_{n-1} * a_n) : (b_1 * \cdots * b_n),$$

    is an composition of all proportions in the arrangement $A_1 \oplus A_2 \oplus \ldots \oplus A_n$, and C is simultaneously an n-level commensurable chain of proportions.

    (d) The **second Composition Formula**: Let $A \cong a : b$ be a number proportion. Here we can assume that a and b are alien to each other, there do not have common prime factors. Then the m-fold composition of the proportion A is the m-step proportion chain and the composition - respective: a similar example of this chain - can very easy calculated like this formula:

$$(\oplus A)^m = A \oplus A \oplus \cdots \oplus A$$

$$\cong a^m : a^{m-1}b : a^{m-2}b^2 : \ldots : ab^{m-1} : b^m.$$

    This represents the stratification, formulated as a chain of proportions, of a musical interval $[a, b]$ corresponding to the given proportion A.

3. Connectable Proportion-Chains

    (a) All real proportion chains A and B are connectable to $A \oplus B$ and to $B \oplus A$.

    (b) All commensurable chains A and B are connectable to $A \oplus B$ and to $B \oplus A$.

    (c) All proportion chains which are similar to number proportion chains are connectable.

Proof First we show statement (1.) and start with implication (a) $\Rightarrow$ (c):
Let us suppose we have a composition and thus an $(n + m)$-level chain

$$C = c_0 : c_1 : \ldots : c_n : c_{n+1} : \ldots : c_{n+m}.$$

Then the front partial-chain

$$A' = a'_0 : a'_1 : \cdots : a'_n = c_0 : c_1 : \cdots : c_n$$

of C is similar to A, and the trailing subchain

$$B' = b'_0 : b'_1 : \cdots : b'_m = c_n : c_{n+1} : \cdots : c_{n+m}$$

of C is similar to B, and apparently

$$a'_n : b'_0 = c_n : c_n \cong 1 : 1,$$

so that the indirect connectivity of B with A is shown. The implication (c) $\Rightarrow$ (a) now results
as follows: Let thus be

$$A' = a'_0 : a'_1 : \ldots : a'_n \cong A \quad \text{and} \quad B' = b'_0 : b'_1 : \ldots : b'_m \cong B$$

two proportion chains that satisfy the composition condition $a'_n : b'_0 \cong 1 : 1$. We can also
write this condition in the form $a'_n : 1 \cong b'_0 : 1$ using the cross rule. But then according to
the exchange rule of Theorem 1.3 the following can happen

$$a'_n : x \cong b'_0 : x \quad \text{as well as} \quad x : a'_n \cong x : b'_0$$

applies to all magnitudes x that enter into a proportion with a'_n and b'_0. In particular

$$a'_{n-1} : a'_n \cong a'_{n-1} : b'_0.$$

Therefore, $A'' = a'_0 : a'_1 : \ldots : a'_{n-1} : b'_0$ is an n-step proportion chain similar to A. By
definition, however, A'' is connectable with B', since they share the connecting member.
Thus the $(n + m)$-step chain is

$$C = a'_0 : a'_1 : \ldots : a'_{n-1} : b'_0 : b'_1 : \ldots : b'_m$$

the desired composition.

From (c) follows (b), since case (2.) with $p = q = 1$ is a special case of (4.).

Now we show the implication (b) $\Rightarrow$ (c) and at the same time introduce the **'extension procedure'** used for this purpose:

Let $A' = a'_0 : a'_1 : \ldots : a'_n$ and $B' = b'_0 : b'_1 : \ldots : b'_m$ be given proportion chains with the commensurable proportion of the terminal magnitudes

$$a'_n : b'_0 \cong p : q \text{ with positive numbers } p \text{ and } q.$$

Then we **expand** A' with q and B' with p – which means that all magnitudes a'_k are multiplied by q and all magnitudes b'_k are multiplied by p – which does not change anything about all step proportions of A' and B'. The new chains A' and B'

$$A'' = a''_0 : a''_1 : \ldots : a''_n \text{ and } B'' = b''_0 : b''_1 : \ldots : b''_m$$

are thus similar to A' and B' and thus similar to A and B As a result we achieve the equality proportion with the multiplication rule as follows:

$$a'_n : b'_0 \cong p : q \Longleftrightarrow q * a'_n : p * b'_0 \cong p * q : q * p \cong 1 : 1.$$

Therefore, a''_n and b''_0 apparently satisfy the exchange condition

$$a''_n : b''_0 \cong q * a'_n : p * b'_0 \cong 1 : 1,$$

so that the indirect connectivity for B with A is realized.

Now to statement (2): The proof consists in an 'inductive' application of this extension procedure. To make this argument more clear we first give two simple examples. It has to be mentioned that we already adapt the form of the numbering to the general case:

Let $A_1 = a_1 : b_1$ and $A_2 = a_2 : b_2$ be two commensurable proportions. Then both are similar to the number-chains, according to the definition.

$$A'_1 = n_1 : m_1 \text{ and } A'_2 = n_2 : m_2 \text{ with positive numbers } n_1, n_2, m_1, m_2.$$

Now we extend A'_1 with n_2 and A'_2 with m_1, so that two new proportions

$$A''_1 = n_1 n_2 : m_1 n_2 \text{ and } A''_2 = n_2 m_1 : m_1 m_2$$

occur, whose connection magnitudes are equal. Since extending proportions leads to similar proportions, these new chains are similar to the initial proportions A_1 and A_2. Thus, a 2-step chain formation is possible and

$$C = n_1 n_2 : m_1 n_2 : m_1 m_2$$

is the composition of both commensurable proportions to a chain, which is also commensurable.

Now we treat the case of three proportions in similar detail, because it gives us a proper insight into a general methodological composition procedure:

Let $A_1 = a_1 : b_1, A_2 = a_2 : b_2$ and $A_3 = a_3 : b_3$ be three commensurable proportions. Then they each are similar to the chains, according to the definition.

$$A'_n = n_1 : m_1, A'_2 = n_2 : m_2 \text{ and } A'_3 = n_3 : m_3$$

with positive numbers $n_1, n_2, n_3, m_1, m_2, m_3$. Analogous to the case of two proportions just described we add the first two proportions A'_1 and A'_2 to the 2-step chain

$$C = n_1 n_2 : m_1 n_2 : m_1 m_2$$

together and then we fulfill the task to add the third chain $A'_3 = n_3 : m_3$ to this chain C to a then 3-level chain afterwards. This is not difficult and works according to the same pattern: We extend this chain C with n_3 and the proportion A'_3 with $m_1 m_2$. Then we get the following two chains

$$C' = n_1 n_2 n_3 : m_1 n_2 n_3 : m_1 m_2 n_3 \text{ and } A''_3 = n_3 m_1 m_2 : m_1 m_2 m_3,$$

which are -again- similar to their predecessors and for which the connection magnitudes are identical. Therefore the proportion chain

$$C'' = n_1 n_2 n_3 : m_1 n_2 n_3 : m_1 m_2 n_3 : m_1 m_2 m_3$$

is the desired 3-step composition of all three given proportions.

Although we could even give a memorable formula for this procedure – and we intend to do so shortly – we still want to finish the induction argument for a general case: To do this, we show that we can always attach a commensurable proportion (and more generally: another commensurable number-proportion chain) to a commensurable n-step number-chain.

Let $A = a_0 : a_1 : \ldots : a_n$ be an n-level commensurable proportion chain, and let $B = b_0 : b_1$ be another commensurable proportion. Now we assume for A that an n-level commensurable number proportion-chain $A' = n_0 : n_1 : \ldots : n_n$, which is similar to A, has already been found. Then let $B' = m_0 : m_1$ be a commensurable number proportion similar to B. The inductive argument now consists in showing that we can add B' to A' to an $(n + 1)$-level chain.

To this end, we extend A' with m_0 and B' with n_n and obtain the commensurable-chains that are also similar to the respective initial-chains

$$A'' = m_0 n_0 : m_0 n_1 : \cdots : m_0 n_n \text{ and } B'' = n_n m_0 : n_n m_1,$$

which again have the same connection magnitudes. Therefore the $(n + 1)$-step proportion chain

$$C = m_0 n_0 : m_0 n_1 : \ldots : m_0 n_n : n_n m_1$$

is both an composition of B to A and simultaneously an composition of all $(n + 1)$ commensurable proportions

$$A_1 = a_0 : a_1, A_2 = a_1 : a_2, \ldots, A_n = a_{n-1} : a_n \text{ and } B = b_0 : b_1$$

to a $(n + 1)$-step commensurable proportion chain

$$C \cong A_1 \oplus A_2 \oplus \ldots \oplus A_n \oplus B.$$

In the end the final conclusion is another formulation of the construction just shown: If an n-step commensurable chain is given, this is equivalent to the fact that n commensurable (step) proportions are given. In turn these each have – by definition – n similar number proportions, which we can then join together in the intended sequence. The result is similar to the given proportion-chain, since all proportions together with their sequence are preserved.

Finally, the given formula (first composition formula) is a copy of our described procedure and one can easily convince oneself that the k-th proportion step A'_k of the given proportion formula has exactly the proportion $n_k : m_k$, i.e. that we have the similarities

$$A'_k \cong A_k \cong n_k : m_k \text{ for } k = 1, \ldots, n.$$

We can easily read this from the concrete products of the proportions, since all other factors of the respective magnitudes of A'_k coincide and thus – using the multiplication rule – shorten out again: It is

$$(n_1 * n_2 * \cdots * n_n) : (m_1 * n_2 * \cdots * n_n) \cong n_1 : m_1,$$
$$(m_1 * n_2 * \cdots * n_n) : (m_1 * m_2 * n_3 \cdots * n_n) \cong n_2 : m_2, \ldots,$$
$$(m_1 * m_2 * \cdots * m_{n-1} * n_n) : (m_1 * \cdots * m_n) \cong n_n : m_n.$$

The 2nd composition formula can be obtained as a repeated application of the 1st composition formula. Similarly, the group of statements (3.) is a consequence of (2.) and the extension technique for number proportion chains embodied in it with respect to their composition. Thus our theorem is illuminated in its justification.

Examples of the Chain Composition Method
With examples we descripe the procedure of concretely adjointing two given chains. From a methodological point of view, we can list three cases of how to unite (merge) two chains into one composition:

  I. The case of real number proportion chains,
  II. The case of rational (commensurable) number proportion-chains,
  III. The case of commensurable-connectable proportion chains.

Case I: Real Number Proportion-Chains Given are the two proportion chains

$$A = 2 : 3 : \sqrt{12} \text{ and } B = \sqrt{3} : 11 : 13.$$

We are to append B to A. To do this, we extend B by the factor $\lambda = \sqrt{12}/\sqrt{3} = \sqrt{4} = 2$ and obtain with $2\sqrt{3} = \sqrt{12}$ the chain similar to B

$$B' = \sqrt{12} : 22 : 26,$$

which by definition can be added directly to A, so that the 4-step proportion chain

$$C = A \oplus B = 2 : 3 : \sqrt{12} : 22 : 26$$

represents a desired composition of A and B. In this case, we could also choose the reverse order. After that we would extend the now 'second' chain A with the factor $\lambda = 13/2$ in order to get the chain similar to A

$$A' = 13 : 39/2 : (13/2)\sqrt{12},$$

which can be added directly to B, and the 4-step composition would result

$$D = B \oplus A = \sqrt{3} : 11 : 13 : 39/2 : (13/2)\sqrt{12},$$

which we see for certain cannot be similar to the chain $C = A \oplus B$.

Case II: Rational (Commensurable) Number-Proportion-Chains We copy the preceding procedure, but ensure by a two-sided extension that the chains retain their integers when given in this form. If two chains A and B are

$$A = a_0 : a_1 : \ldots : a_n \text{ and } B = b_0 : b_1 : \ldots : b_m$$

with rational numbers a_j, b_k, then we ensure by extensions of A with the main denominator λ of all numbers a_j and of B with the main denominator μ of all numbers b_k, that the new chains A' and B', which are similar to the old chains, are integer. Therefore, we can – a priori – assume that both chains A and B *are* in integer form at this time. Then we make the composition as in the following example: Given

$$A = 3 : 4 : 7 \text{ and } B = 2 : 5 : 6 : 11.$$

Then we extend the chain A by 2, the chain B by 7, and get the similar chains

$$A' = 6 : 8 : 14 \text{ and } B' = 14 : 35 : 4 : 77,$$

which now form a 5-step composition

$$C = A \oplus B = 6 : 8 : 14 : 35 : 42 : 77,$$

and they are merged.

Case III: Commensurable-Connectable Proportion Chains Let

$$A = a_0 : a_1 : a_2 \text{ and } B = b_0 : b_1$$

and let the proportion of the magnitudes connecting them be $a_2 : b_0 \cong 3 : 4$ and $a_2 : 3 \cong b_0 : 4$, respectively. Then we expand the chain A by 4 and the chain B by 3 and obtain the following two chains

$$A' = a_0' : a_1' : a_2' = 4a_0 : 4a_1 : 4a_2 \text{ and } B' = b_0' : b_1' = 3b_0 : 3b_1,$$

which are obviously similar to A *and* B. Now applied to the new proportions to be connected it can be followed:

$$a_2' : b_0' \cong 4a_2 : 3b_0 \cong (4 * 3) : (3 * 4) = 12 : 12 \cong 1 : 1,$$

so that we have the two mutually similar and 'almost' identical 3-level-chains

$$C = A \oplus B = 4a_0 : 4a_1 : 4a_2 : 3b_1 \text{ or } C = A \oplus B = 4a_0 : 4a_1 : 3b_0 : 3b_1$$

as compositions. Depending on the kind of magnitudes we cannot write $4a_2 = 3b_0$ – except the trivial case of number-proportions. Nevertheless, it is clear to see that it makes some things easier (at least from a strictly formal point of view).

The 2nd composition formula is used in particular when a fixed musical interval is constantly iterated. Exemplary when fifth is set upon fifth for the purpose of scale generation and this process then continues for a while. A numerical example may illustrate this:

Example 2.4 Proportion Chain of an Interval Layering

We place 6 Pythagorean whole tones (8 : 9) on top of each other.

Question: What is the corresponding proportion chain?

Answer: Because of the divisor-unrelated situation, the form of this resulting 7-member-chain that cannot be simplified further is the number-proportion chain

$$8^6 : 8^5 9 : 8^4 9^2 : 8^3 9^3 : 8^2 9^4 : 8^1 9^5 : 9^6$$
$$= 262144 : 294912 : 331776 : 373248 : 419904 : 472392 : 531441.$$

The outer proportion 262144 : 531441 is only almost as large as 1 : 2; strictly speaking, we have the difference to the octave 1 : 2 in the proportion 524288 : 531441, the small 'error interval' of Pythagorean tuning – generally known as the **'Pythagorean comma'** and of significant influence on the entirety of musical tone-theory. We have already encountered it in Example (1.2) of Sect. 1.4. ◀

We come to another important rule that is used at every turn in musical practice:

The Octave Principle in the Practice of Instrument Tuning

Suppose we were about to tune an organ or piano according to a certain desired temperament – equal temperament, mean-tone temperament, Kirnberger III, and so on. Then the usual procedure can propably be described as follows:

We first tune a single octave scale – that is, usually the 12 notes of the keyboard – say from the note $A_1{\sim}440$ Hz to its octave $A_2{\sim}880$ Hz inclusive. Mathematically, this means that we take the given 12-step proportion chain

$$A = a_0 : a_1 : \ldots : a_{12},$$

whose stage proportions correspond to the required frequency factors of the desired temperature control and where the total proportion of A with

$$a_0 : a_{12} \cong 1 : 2$$

represents the octave, as a model of the octave scale.

Once this one octave has been tuned as desired, we will certainly refrain from copying this process – which is not infrequently laborious – for all the octaves that follow upwards or downwards. Rather, we will exemplary set the following note B_2, which is a semitone step above the already tuned A_2, as a pure octave to B_1, which we have set up in the first step. And so it goes on, step by step, note by note.

Conclusion: *If we have tuned a single octave correctly according to the given wish, we have – in a flash – tuned the whole instrument.*

That this works out is beyond question, as it has been tested millions of times. But what interests us at this point is the question of the mathematical principle behind it! We now want to describe this principle in general – but also give it the name of the method just explained:

Theorem 2.2 (The Octave Principle)

Let be given the two connectable n-step proportion chains

$$A = a_0 : a_1 : \ldots : a_n \text{ and } B = b_0 : b_1 : \ldots : b_n,$$

and let it be the 2n-step proportion chain

$$C = A \oplus B \cong c_0 : c_1 : \ldots : c_n : c_{n+1} : c_{n+2} : \ldots : c_{2n}$$

whose composition is $A \oplus B$. Then the following conditions are equivalent:

1. $A \cong B$.
2. All $(n + 1)$ proportions $c_k : c_{k+n}$, $(k = 0, \ldots, n)$ are similar to each other and thus similar to the (starting) proportion $c_0 : c_n \cong a_0 : a_n$, the total proportion of A. Expressed in mathematical formalism, this means:

$$c_k : c_{k+n} \cong c_j : c_{j+n} \cong a_0 : a_n \text{ for all } j, k = 0, \ldots, n.$$

In particular, $b_0 : b_n \cong c_n : c_{2n} \cong a_0 : a_n$.

Conclusion (The Octave Principle in Instrument Tuning Practice)

If $A = a_0 : a_1 : \ldots : a_n$ is a given n-step proportion chain that we can adjoint with itself. Then the proportion-chain T, obtained by composition continued arbitrarily often on both sides

$$T = \cdots \oplus A \oplus A \oplus A \oplus \ldots$$

is periodic – that is, a shift of n (up or down) steps always gives the same total proportion of A (or its inverse):

(continued)

> **Theorem 2.2** (continued)
>
> If $T = \cdots : t_k : t_{k+1} \cdots$, then
>
> $t_k : t_{k+n} \cong a_0 : a_n$ for all tones k of the iteration – chain T, respectively
>
> $t_k : t_{k-n} \cong a_n : a_0$ for all tones k of the iteration – chain T.
>
> Thus T is a periodically tuned **keyboard.** The standard case would be given by $n = 12$, and this then describes our usual piano/organ/keyboard.

We want to prove this 'musically' – because we succeed as follows:

▶ **Important**

We can prove this theorem in exactly the same way as we would tune the piano: Indeed, this is exactly what lies behind the following recursive reasoning, which we start with the first stage $k = 1$: After defining the composition-chain, we have the similarities

$$c_0 : c_1 \cong a_0 : a_1 \quad \text{and} \quad c_n : c_{n+1} \cong b_0 : b_1.$$

Therefore, $c_0 : c_1$ and $c_n : c_{n+1}$ are similar exactly when this applies to $a_0 : a_1$ and $b_0 : b_1$. But then the principle of the 'pure octave' follows with the cross-rule:

$$c_0 : c_1 \cong c_n : c_{n+1} \Leftrightarrow c_0 : c_n \cong c_1 : c_{n+1}.$$

And because $c_0 : c_n \cong a_0 : a_n$ is satisfied, it follows (doing the same for each subsequent stage) the assertion.

The introductory description of the tuning procedure of an instrument is 'proved' by the following sentence: For $n = 12$, all positions k and $k + 12$ are octaves and then if the initial scale is a 12-step octave scale, $a_0 : a_{12} \cong 1 : 2$, and all pairs of keys $(k, k + 12)$, $k = 0, \ldots,$ 12 then have this exact proportion. Furthermore, if the direction of play and the octave proportion are reversed, so do all down octaves – that is, all conceivable keyboards up and down. Figure 2.2 illustrates this with the keyboard.

Of course we would like to note that this 'principle' merely describes – as the name suggests – a basic method, a basis, for keyboard instrument tuning, professional, artistically designed tuning takes in addition some others completely different approachs,

Octave tuning **principle**: We only need to tune the 12 notes of a single chromatic
octave scale (e.g. a **reference octave** c_0 - h_0 , dotted) - then **all other notes of** the
entire keyboard are to be tuned octave-pure by means of easy-to-perform octave
tuning from the corresponding notes of the reference octave - done.

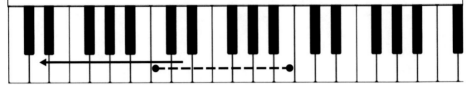

Fig. 2.2 The octave principle of tuning

for it is only through chromatic major chord chains, desired and necessary beats of almost
hidden sounding harmonics and a few other mysterious ingredients that the masters of their
craft find the way to the true euphony of the instrument. (It is said that the great Vladimir
Horowitz never did without his own piano tuner as a constant companion...).

We conclude this consideration of this important basic construction of compositions with
an initial listing of mathematical structural laws that accompany these processes:

Theorem 2.3 (The Algebra of the Composition Operation)
1. **Invariance under Similarity:** The composition always refers to whole similarity
   classes: If A and B are connectable to the chain $C \cong A \oplus B$, then also all chains A'
   and B' similar to A and to B respectively are connectable and their composition is
   similar to the composition C, thus symbolically applies

$$A' \cong A \ \text{and} \ B' \cong B \Rightarrow A' \oplus B' \cong A \oplus B$$

2. **No commutativity:** For composition, the law of commutativity – that is,
   interchangeability – is generally neither true nor meaningful:

$$A \oplus B \ \text{and} \ B \oplus A$$

need not be similar and usually are not – even if both forms were constructible.
Exceptions would only be special symmetrical constellations such as $B \cong A$ and
similar others.
   However, for the case of two chains of equal length A and B (i.e. for $n = m$), the
more precise description is indeed valid

(continued)

Theorem 2.3 (continued)
$$A \oplus B \cong B \oplus A \Leftrightarrow B \cong A.$$

Moreover, for chains of also equal length, the above statement (1) about invariance under similarity can also be used to prove the equivalence

$$A \oplus B \cong B \oplus C \Leftrightarrow A \cong B \cong C$$

assume that all these compositions exist at all.

3. **Associativity and Transitivity:** If C is connectable to B and B is connectable to A, then the following transfer mechanisms holds without restriction:

   1. **Associativity:** both $B \oplus C$ is connectable to A and C is connectable to $A \oplus B$, and then both compositions are the same (more precisely: similar):

   $$A \oplus (B \oplus C) \cong (A \oplus B) \oplus C.$$

   2. **Transitivity:** C is then also connectable to A, so $A \oplus C$ exists.

4. **Inverse:** For all proportion chains A and B yields: If B is connectable to A, then A^{inv} is connectable to B^{inv}, and the rule of arithmetic applies:

$$(A \oplus B)^{inv} \cong B^{inv} \oplus A^{inv}.$$

Why? The first statement follows directly from the definition: A chain $C = A \oplus B$ has exactly the same sequence of step proportions as A (front chain) and B (back chain), and thus this also applies to all chains similar to A *and* B, respectively, since their sequence of step-proportions is identical – that is, similar.

Also the other formulas and statements can be won without much effort. Only the last rule (4.) we want to prove once in detail in its generality for the exercise, and for this let be

$$A = a_0 : a_1 : \ldots : a_n \text{ and } B = b_0 : b_1 : \ldots : b_m$$

given proportion chains, and the chain

$$C = A \oplus B = c_0 : c_1 : \ldots : c_n : \ldots : c_{n+m-1} : c_{n+m}$$

be its composition. By definition, the inverse proportion chain to C is the chain

$$C^{inv} = c_{n+m} : c_{n+m-1} : \ldots : c_n : \ldots : c_1 : c_0.$$

Now, according to our definition of an composition $A \oplus B$ the similarities are valid

$$c_0 : c_1 \cong a_0 : a_1, \ldots, c_{n-1} : c_n \cong a_{n-1} : a_n$$
$$c_n : c_{n+1} \cong b_0 : b_1, \ldots, c_{n+m-1} : c_{n+m} \cong b_{m-1} : b_m.$$

We can define these according to the inverse rule as a sequence of proportions

$$c_{n+m} : c_{n+m-1} \cong b_m : b_{m-1}, \ldots, c_{n+1} : c_n \cong b_1 : b_0$$
$$c_n : c_{n-1} \cong a_n : a_{n-1}, \ldots, c_1 : c_0 \cong a_1 : a_0$$

so that likewise by definition the entire chain

$$c_{n+m} : c_{n+m-1} : \ldots : c_n : \ldots : c_1 : c_0$$

is an composition of A^{inv} to B^{inv}. This confirms our rule.

2.3   Reciprocal Proportion Chains

One of the most interesting aspects of proportion chain-theory arises when we simply reverse the sequence of **step proportions**. Let's take an example:

The 2-step proportion chain $4 : 5 : 6$ stands for the diatonic pure major chord; the pure major third ($4 : 5$) is followed by the pure minor third ($5 : 6$), and the span is a fifth ($4 : 6 \cong 2 : 3$). If we reverse the order, we as musicians know what comes out: a minor chord. Now, to construct the 2-step proportion chain as an composition of (first) minor and (then) major third, we apply the appropriate expansion so that the old connecting magnitudes (6 and 4) become equal; the 'kgV' ("kleinstes gemeinsames Vielfaches" (in german) - the least common multiple) is 12 – that is to say

$$5 : 6 \cong 10 : 12 \text{ and } 4 : 5 \cong 12 : 15 \Rightarrow (5 : 6) \oplus (4 : 5) \cong 10 : 12 : 15,$$

and so the proportion chain of the minor triad is created.

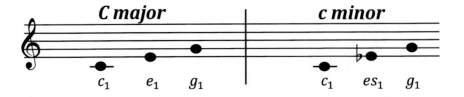

On the other hand, once we form the proportion chain of the 'reciprocals' of the initial chain, if we order by size, we get the following proportion chain:

$$\frac{1}{6} : \frac{1}{5} : \frac{1}{4} \cong \frac{6 * 5 * 4}{6} : \frac{6 * 5 * 4}{5} : \frac{6 * 5 * 4}{4} \cong 20 : 24 : 30 \cong 10 : 12 : 15,$$

thus the same chain of proportions as before, when we simply reversed the order of the proportions. This highly remarkable connection but on the other hand it is certainly computationally easy to see through. But it can be said that it is nevertheless extremely profoundly embedded in the concept of chains of proportions and their music. We formulate in advance:

▶ The proportion chain of the ordered reciprocals of the magnitudes of an ordered (number-) proportion chain is itself a chain in which the order of the proportions is reversed.

At least that is what we saw in the example. Certainly this interesting interplay between reciprocals and reversed proportions is also valid in general and interesting symmetries come to light, namely when certain mean value relations come into play. However, we will deal with this in the following chapter. We start with the definition:

Definition 2.5 (Reversible, Reciprocal Proportion Chains)

An n-step proportion chain $A = a_0 : a_1 : \ldots : a_n$ is called **reversible** if there is an also n-step proportion chain

$$B = b_0 : b_1 : \ldots : b_n$$

whose step proportions are similar to the step proportions of A listed in reverse order, i.e. if

$$b_0 : b_1 \cong a_{n-1} : a_n, \ \ b_1 : b_2 \cong a_{n-2} : a_{n-1}, \ \ldots, b_{n-1} : b_n \cong a_0 : a_1$$

holds. Every such proportion chain B is called a **reciprocal of A**, and we use for this (i.e. for the totality of all reciprocals of the given proportion chain A) the notation A^{rez}. In mathematical notation it reads like this:

$$A^{\text{rez}} = \{B \mid B \text{ is an n} - \text{step proportion chain which is reciprocal to } A \}.$$

Comments

1. We will suppress the distinction of the symbol A^{rez} as a single chain of proportions – reciprocal to A – or as the totality of all such chains in favour of a better readability. Anyway, all rules and calculations do not really take place in the single chains of proportions, but in their complete similarity classes. This is supported in the following remark (2.):

2. Because only the 'ratios' (proportions) of all magnitudes (and not the measure of the magnitudes themselves) are important in the definitional specification, it is quickly clear that similar chains also have similar properties (keyword 'invariance under similarity'). In particular, we recognize the following generalizations from the definition of 'reciprocal':

   (a) If a proportion chain $\overline{K}$ is reciprocal to a chain K, then so are all chains which are themselves similar to the reciprocal.

   (b) If a proportion chain $\overline{K}$ is reciprocal to a chain K then it is also reciprocal to all chains similar to the chain K.

   After all, the proportions of all stages do not change under similarity.

3. Other names for **'reciprocal proportion chain'** would also be, for example, 'a proportion chain **mirrored to** the given chain' or 'an **inverse proportion chain**'. Both denominations correspond to obvious constructive contexts but we have to pay attention to the fact that **'reciprocal'** and **'inverse'** always differ – except for the case of constant equal magnitudes; therefore 'inverted proportion chain' would not be a safe denotation either, in short:

▶ **'Reciprocal'** means that the **step proportions** are adjoined in reverse order; **'inverse'**, on the other hand, means that the members **(magnitudes) are** written in reverse order, and then all the step proportions are merely inverted and adjoined in reverse order – so this is comparatively equivalent to 'reading backwards' the given chain.

To give an explanatory example from simple chordal-theory: The proportion chain of the major chord, already considered in our section introduction, is given by the chain

$$A = 4 : 5 : 6 - \text{interpretable as : major third upwards} \oplus \text{minor third upwards.}$$

If the magnitude were merely reversed, the chain would be

$$A^{inv} = 6 : 5 : 4 - \text{interpretable as : minor third downwards} \oplus \text{major third downwards,}$$

which describes the proportions of the same major chord, but from the downward point of view of the final note of A, the dominant. 'Reciprocal', on the other hand, means the reverse order of the proportions of the chain, especially not its magnitudes. Applied to this major triad $4 : 5 : 6$, its reciprocal means the sequence of first a minor third ($5 : 6$) and then a major third ($4 : 5$), which we know as a minor triad, thus resulting, exactly as described in the introduction – in the following situation

$$A^{\text{rez}} = (5 : 6) \oplus (4 : 5) \cong 10 : 12 : 15.$$

By the way, in contrast to the reciprocal, an inverse proportion chain can be indicated by the same magnitudes. In the case of the reciprocal, on the other hand, the appending process must still be included by suitable similarity operations.

4. A reciprocal to a 1-step proportion chain $A = a : b$ is the chain A itself (and exactly all other proportions similar to it), in mathematical symbolism:

$$A = a : b \Leftrightarrow A^{\text{rez}} \cong a : b, \quad \text{whereas } A^{inv} = b : a.$$

Whether there is a reciprocal-chain at all for a given proportion chain is not clear a priori; we will find corresponding statements on this in the coming theorem, as well as bundle other essential mathematical facts around the term 'reciprocal' into a package:

Theorem 2.4 (Existence, Calculation and Symmetries of Reciprocal Proportion Chains)

1. **Invariance under similarity:**
   If B is reciprocal to A, then all chains of proportions similar to B are reciprocal to all chains of proportions similar to A; in mathematical language:

   $$B \cong A^{\text{rez}} \Leftrightarrow \bar{B} \cong \bar{A}^{\text{rez}} \text{ for all chains } \bar{A} \text{ and } \bar{B} \text{ with } \bar{A} \cong A \text{ and } \bar{B} \cong B.$$

2. **Symmetries:**
   (a) It always applies $B \cong A^{\text{rez}} \Leftrightarrow A \cong B^{\text{rez}}$.
   (b) For the double reciprocal: $(A^{\text{rez}})^{\text{rez}} \cong A$.
3. **Calculation formulas (reciprocal formula):**
   If $A = a_0 : a_1 : \ldots : a_n$ is a concrete number proportion-chain, then

---

(continued)

Theorem 2.4 (continued)
$$B = b_0 : b_1 : \cdots : b_m \cong \frac{1}{a_n} : \frac{1}{a_{n-1}} : \cdots : \frac{1}{a_0}$$

is a reciprocal to A. By extension of this chain B with the 'main denominator'

$$\mu = a_0 * a_1 * \cdots * a_n \equiv a_0\, a_1 \ldots a_n$$

i.e. the product of all their magnitudes, one reaches the formally fraction-free form of a reciprocal, which we characterize as a reciprocal formula:

$$A^{\text{rez}} \cong (a_0 \ldots a_{n-1}) : (a_0 \ldots a_{n-2}\, a_n) : \cdots : (a_0\, a_2 \ldots a_n) : (a_1 \ldots a_n).$$

4. **Existence of reciprocals:**
   Every n-step number proportion-chain has a reciprocal and it follows therefore the following:
   (a) Every chain similar to a number proportion-chain has a reciprocal.
   (b) Every commensurable proportion chain has a reciprocal, and all reciprocals are also commensurable again.

Remark on the Reciprocal Formula

The formula for this fraction-free form of a reciprocal is easy to remember, despite its possibly daunting length. Let us use the expressions

$$A^{\text{rez}} \cong \mu_0 : \cdots : \mu_n \text{ and } \mu = a_0 * \cdots * a_n,$$

this results in a product structure that is easy to remember: From the total product μ of all magnitudes is missing

- in the first magnitude $\mu_0 = a_0 * \ldots * a_{n-1}$ the factor a_n,
- in the second magnitude $\mu_1 = a_0 * \ldots * a_{n-2} * a_n$ the factor a_{n-1},
- $\ldots$,
- in the last magnitude $\mu_n = a_1 * \ldots * a_n$ the factor a_0.

In other words, let's put to the abbreviations $\mu_k = \mu / a_{n-k}$ (for $k = 0, \ldots, n$), ie.

$$\mu_0 = \frac{\mu}{a_n}, \mu_1 = \frac{\mu}{a_{n-1}}, \ldots, \mu_{n-1} = \frac{\mu}{a_1}, \mu_n = \frac{\mu}{a_0},$$

so the formal fraction-free form of a reciprocal is the n-step proportion chain

$$A^{\text{rez}} \cong \mu_0 : \ldots : \mu_n$$

with the magnitudes $\mu_0, \ldots, \mu_n$, which in the case of positive integer magnitudes a_k are themselves integer again and thus form a commensurable proportion chain.

Proof of the theorem: First to the statement (1.) and (2.): According to our definition of a reciprocal as an (arbitrary n-step) chain of proportions whose step-proportions are in reverse order, these statements are in any case immediately clear. The two chains given in (3.) satisfy the proportion specifications of reciprocals. This can be easily shown by a short check, such as simple arithmetic:

$$\frac{1}{a_{k+1}} : \frac{1}{a_k} = \frac{a_k}{a_k a_{k+1}} : \frac{a_{k+1}}{a_k a_{k+1}} = \frac{a_k}{\mu} : \frac{a_{k+1}}{\mu} \cong a_k : a_{k+1}.$$

If one writes down the reciprocals in reverse order, the step proportions similar to the chain A arise – but in the mirrored order. A common extension factor (μ) for all magnitudes does not change the proportions – therefore the formal fraction-free form (the 'reciprocal formula') is also a similar reciprocal.

Now the existence (statement 4.) results as a consequence from this reciprocal formula. Both conclusions are proved too, because according to Theorem 2.1 every commensurable proportion-chain is similar to a number-proportion-chain and considering the invariance under similarity then its reciprocal is also a reciprocal to the given chain. As all step-proportions of a reciprocal are preserved (they are only arranged inversely), they are also commensurable, so the theorem follows.

Example 2.5 Proportion Chains and Their Reciprocals

No.	Proportion chain A	One of their reciprocals A^{rez}
1	$3 : 4$	$3 : 4$
2	$3 : 4 : 5$	$12 : 15 : 20$
3	$1 : 2 : 3 : 4 : 5 : 6$	$10 : 12 : 15 : 20 : 30 : 60$
4	$3 : 5 : 7 : 9$	$35 : 45 : 63 : 105$
5	$1 : 3 : 3 : 3 : 6$	$1 : 2 : 2 : 2 : 6$
6	$30 : 36 : 40 : 45 : 50 : 60$	$30 : 36 : 40 : 45 : 50 : 60$

◄

In Examples 2.2, 2.3, and 2.4, the proportion chain A is arithmetic. Its reciprocal A^{rez} is harmonic – we will explain this in later Sects. 3.3 and 3.4; the last example shows us a 'symmetrical' proportion chain; the proportions of diametrically located steps are similar – we will examine this important structure in more detail in the next Sect. 2.4. With this in mind, we hold for now:

▶ The interplay of the inversion of the stages of a chain of proportions with games of averaging and their common musical interpretations shapes and permeates large parts of the conceptualizations as well as the **theory of historical music**. The culmination of these connections is represented here by the **Harmonia perfecta maxima,** in which a *theory of proportions* – from the point of view of the ancient world – *'perfectly in equilibrium'* becomes one with what was understood as the theory of music of the past millennia.

This meaning of harmonia perfecta maxima runs like a thread through the teachings of the ancients - from the Pythagoreans and the Greek scholars to Boethius and the theorists in the times of antiquity, the Middle Ages and the ealy modern period. This can be seen in sometimes virtuoso interpretations and descriptions. We will learn more about this in Sect. 3.2.

2.4   Symmetrical Proportion Chains

In music-theory, symmetrical chains of proportions play a special role. We will become acquainted with this in abundance in all sections dealing with harmonia perfecta.

Definition 2.6 (Symmetrical Proportion Chains)
A proportion chain is called **symmetrical** if it is reciprocal to itself. This is the case if all the step proportions are similar to the step proportions listed in reverse order. In mathematical language this means the following:

The n−step proportion chain $A = a_0 : a_1 : \ldots : a_n$ is symmetrical

$$\Leftrightarrow a_k : a_{k+1} \cong a_{n-k-1} : a_{n-k} \text{ for all parameters } k = 0, \ldots, n-1.$$

For the special cases of 2- or 3-step proportion chains, this means:

- The 2-step chain $a : b : c$ is symmetrical when $a : b \cong b : c$;
- the 3-step chain $a : b : c : d$ is symmetrical when $a : b \cong c : d$.

Consequently, in the case of 2-step chains, the magnitude b is then the **Geometric Mean** of the outer magnitudes a and c.

In the following theorem we write down some useful criteria for the occurrence of the proportion chain symmetry. Depending on the approach we discover a number of interesting conditions that guarantee the symmetry of proportion chains. Among themselves all these conditions are equivalent, but their applications become efficient just by knowing the range of their representing mechanisms.

Theorem 2.5 (Symmetry Criteria for Proportion Chains)

The following symmetry properties apply to proportion chains $A = a_0 : a_1 : \ldots : a_n$:

1. **Invariance under Similarity**:

   If a proportion chain is symmetrical, so all chaines that are similar to it are symmetrical.

2. **Symmetry Criteria**:
   1. **Step principle**: A is symmetrical $\Leftrightarrow A \cong A^{\text{rez}}$.

      This then means that all mirrored situated step proportions are similar,

      $$a_k : a_{k+1} \cong a_{k+1}^* : a_k^* \text{ for all } k = 0, \ldots, n-1.$$

   2. **Mirror principle**: A is symmetrical $\Leftrightarrow a_0 : a_k \cong a_{n-k} : a_n$ for all $k = 0, \ldots, n$.

      All proportions mirrored to the outer magnitudes are similar. And even more generally: All proportions mirrored to each other are similar, that means:

      $$a_j : a_k \cong a_k^* : a_j^* \text{ for all } k, j = 0, \ldots, n.$$

   3. **Hyperbola principle**: A is symmetrical $\Leftrightarrow$ all magnitude pairs (a_k, a_k^*), $k = 0, \ldots, n$, lie on the same hyperbola

      $$x * y = a_0 * a_n.$$

   The products $a_k * a_k^*$ $(k = 0, \ldots, n)$ of mirrored magnitudes are equal, so they all have the value of the product of the outer magnitudes $a_0 * a_n$.

   4. **Principle of symmetrical reduction**:

      $$A \text{ is symmetrical} \Leftrightarrow (a_0 : a_1 \cong a_{n-1} : a_n) \text{ and } A_1 = a_1 : a_2 : \cdots :$$
      $$a_{n-1} \text{ is symmetrical.}$$

   Thus, if the two outer proportions are similar, and if the remaining 'symmetrically lying inside A and only $(n–2)$-step partial proportion-chain A_1 is symmetrical, then the whole chain is symmetrical.

(continued)

Theorem 2.5 (continued)

5. **Partial Chain Principle, Subchain Principle**: A is symmetric $\Leftrightarrow$ for every (real) partial chain B with $B \subseteq A$ there is a reciprocal with $B^{\text{rez}} \subseteq A$.

Thus, an n-step proportion chain A is symmetric if and only if for every partial proportion chain B of A *there* is also a partial proportion chain C *of A* which is similar to a reciprocal B^{rez} of B.

Addition: One obtains such a reciprocal C, which is formed simultaneously from the magnitudes of A, with the following abstract procedure: Let

$$B = a_{n_0} : a_{n_1} : \ldots : a_{n_k} \text{ with } 0 \leq n_0 < n_1 < \cdots < n_k \leq n$$

a given k-level (ordered) subchaine of A. Then the (ordered) composition of all magnitudes mirrored in A $(a_{n_j}^*)$ to a k-level chain C

$$C = B^{\text{rez}} = a_{n_k}^* : \ldots : a_{n_0}^* = a_{n-n_0} : a_{n-n_1} : \ldots : a_{n-n_k}$$

is such a partial proportion-chain of A reciprocal to B.

Proof The invariance to similarity operations, which we already have often quoted, is already given in the definitions. So let us come to the symmetry criteria. The step-principle follows the definition and describes this then also by means of the mirrored magnitudes. The mirror-principle follows by successive application of the step-principle and we demonstrate this at a concrete situation, by which we bypass the general indexing-writings. This may be unfamiliar to some readers, but the transfer to the so called 'general case' follows the same pattern.

Suppose we had a 4-step proportion chain

$$A = a_0 : a_1 : a_2 : a_3 : a_4.$$

By definition, it is symmetric exactly when the similarities are

$$a_0 : a_1 \cong a_3 : a_4 \text{ and } a_1 : a_2 \cong a_2 : a_3$$

fulfilled. Let us then see by way of example that the mirrored step proportions are also similar. Now are because of

$$a_3 : a_4 = a_1^* : a_0^* \text{ and } a_2 : a_3 = a_2^* : a_1^*$$

already the similarities

$$a_0 : a_1 \cong a_1^* : a_0^* \text{ and } a_1 : a_2 \cong a_2^* : a_1^*$$

given. Therefore, it remains to demonstrate that

$$a_0 : a_3 \cong a_1 : a_4 = a_3^* : a_0^*$$

are similar. We apply the algebraic rules of fusion from Theorem 1.4 and see the assertion using the short but tricky calculus

$$a_0 : a_3 = (a_0 : a_1) \odot (a_1 : a_2) \odot (a_2 : a_3) \cong (a_3 : a_4) \odot (a_1 : a_2) \odot (a_2 : a_3)$$
$$= (a_1 : a_2) \odot (a_2 : a_3) \odot (a_3 : a_4) = a_1 : a_4.$$

With an analogous calculation we can then show the similarity of all diametrical proportions if all mirrored step proportions are similar. Therefore the step principle and the mirror principle are equivalent. The hyperbolic principle follows by fractional arithmetic interpretation of the proportions, and then from the similarities of the diametrical steps the equality of the required products follows, and this leads to the hyperbolic property (see also Theorem 3.1). Finally, the principle of symmetric reduction, which follows directly from the step principle anyway, provides further possibilities for systematically transferring larger chains to smaller numbers of steps. Similarly, we can conveniently identify the partial proportion-chain principle with the general mirroring principle: If

$$B = a_{n_0} : a_{n_1} : \ldots : a_{n_k} \subseteq A,$$

so we see with this symmetry principle that the chain C of ordered mirrored magnitudes

$$C = a_{n_k}^* : \ldots : a_{n_0}^*$$

is reciprocal to B: For example the similarity is

$$a_{n_0} : a_{n_1} \cong a_{n_1}^* : a_{n_0}^*,$$

and likewise the proof of the similarity of all other stages is carried out.

Example 2.6 Symmetrical Proportion Chains

1. All geometrical chains of proportions are symmetrical; a chain is called 'geometrical' if **all** its step proportions are similar. (We will return to this in connection with the

consideration of classical mean values (medieties)). Thus, if $q > 0$ is a parameter, and if a is any real magnitude with $a > 0$, then the n-step proportion chain

$$A = a : aq^1 : aq^2 : \ldots : aq^n$$

is geometrical and symmetrical; the step proportions are all similar to the proportion $(1 : q)$.

2. The example of the special geometrical (i.e. symmetrical) chain

$$A = 1 : 2 : 4 : 8 : 16$$

shows that the reciprocal of a partial-chain need not be unique even in its form as a partial-chain of A; thus the partial proportion chain $B = 2 : 8 : 16$ has the two chains $1 : 2 : 8$ and $2 : 4 : 16$ as reciprocal-chains; both are partial-chains of A. Both chains are of course similar.

Musically, this chain represents a 'chord' consisting of any keynote and four successive octaves above it, for example the sequence

$$c_0 - c_1 - c_2 - c_3 - c_4.$$

3. The **'musical'** chain of proportions

$$A = 30 : 36 : 40 : 45 : 50 : 60$$

is symmetrical. The outer proportions are similar, because $30 : 36 \cong 50 : 60 \cong 5 : 6$, and the remaining inner chain $36 : 40 : 45 : 50$ is symmetrical because its outer proportions are similar and the remaining inner chain of proportions is only the 1-step proportion $40 : 45$, which is inherently symmetrical (principle of symmetrical reduction). This chain is the proportion-chain of the complete diatonic octave canon.

4. We can conveniently create symmetric proportion chains by adjointing to any chain B its reciprocal: Namely, the
   Theorem (Architecture of Symmetrical Proportion-Chains).
   For any m-step proportion-chain B and any proportion P the two chains

$$A \cong B \oplus B^{\text{rez}} \quad \text{and} \quad A \cong B \oplus P \oplus B^{\text{rez}}$$

are symmetrical. Tthe former is a $(2\,m)$-step-chain, and the latter is a $(2\,m + 1)$-step-chain. The reverse is also true: A proportion chain A is symmetrical exactly if it has one

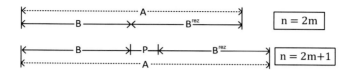

Fig. 2.3 Sketch of the architecture of symmetrical proportion chains

of the above architectural forms – in the case of an even number of steps the former, in the case of an odd number of steps the latter.

◀

To illustrate the architecture of symmetrical-chains of proportions, consider Fig. 2.3.

We will formulate and prove this statement again in the proportion-chain theorem of the next Sect. 2.5. In any case, we can test this theorem about the architecture of symmetric chains of proportions for the time being on the above examples:

First, let us take the chain

$$A = 1 : 2 : 4 : 8 : 16,$$

then the front 'half' partial-chain $1 : 2 : 4$ would be the chain B. Its magnitudes mirrored in A – diametrically – are 16, 8 and 4; the proportion chain $B^* = 4 : 8 : 16$ is symmetrical and thus reciprocal to itself and we get the first of the two architectural formulas.

Second: For the chain of the complete diatonic octave canon

$$A = 30 : 36 : 40 : 45 : 50 : 60$$

$B = 30 : 36 : 40$ is the front partial-chain; its mirrored one is the chain $B^* = 45 : 50 : 60$, and its steps $9 : 10$ and $5 : 6$ are inversely the similar-same as in the case of the chain B:

$$B = (5 : 6) \oplus (9 : 10) \text{ and } B^* = (9 : 10) \oplus (5 : 6) = B^{\text{rez}}.$$

In combination with the both connecting middle proportion $P = 40 : 45$ we also realized the other architectural form.

Third, we finally demonstrate the reverse way: we give ourselves an arbitrary m-step proportion chain B and want to use it to obtain a $2\,m$-step symmetric chain according to the first architectural form:

For example, let $B = 2 : 5 : 6 : 7$. Then we need to construct a reciprocal:

$$B^{\text{rez}} = (6 : 7) \oplus (5 : 6) \oplus (2 : 5) \cong 30 : 35 : 42 : 105.$$

If we connect this to chain B (by extending both chains so that the connection magnitudes have the value $30 * 7 = 210$), the whole 6-step chain A is

$$A \cong \quad (2:5:6:7) \quad \oplus \quad (30:35:42:105)$$
$$\cong \quad 60:150:180:210:245:294:735,$$

whose symmetry on the basis of the step criterion (though somewhat laborious) is nevertheless confirmed by elementary calculation if desired.

A given middle proportion in the form $P = 7 : 30$, on the other hand would immediately connect $B \oplus P \oplus B^{\text{rez}}$ to a 7-step chain

$$A = 2 : 5 : 6 : 7 : 30 : 35 : 42 : 105,$$

which would also provide us with a constructive example of the second architectural form.

2.5   The Proportion Chain Theorem

In this section we summarize (as a mathematical summary of this chapter, so to speak) the most important facts concerning the topic 'chains of proportions' and furthermore connect some statements among each other. We also strive to keep the discussion as general as possible (this is, after all, customary in mathematics). Nevertheless, we demand that the choice of descriptive language should certainly also take this generality into account. This succeeds quite excellently by means of the mathematical alphabet, which is the language of set theory.

> **Theorem 2.6 (Proportion-Chain- Theorem)**
> In the following, let A, B, etc. be general n- or m-level proportion chains. Then there are the following properties for calculating with proportion chains:
>
> 1. **Similarity class structure of all proportion-chains**
>   The similarity is an **equivalence relation** on the totality $\mathcal{M}$ of all n-level proportion chains; from the three – characterizing this structure – properties
>   (a) **Reflexivity:** Every chain is similar to itself: $A \cong A$,
>   (b) **Symmetry:** If B is similar to A, then A is also similar to B: $A \cong B \Leftrightarrow B \cong A$,
>   (c) **Transitivity:** If A is similar to B, B is similar to C, then A is also similar to C:

(continued)

Theorem 2.6 (continued)
$$A \cong B \text{ and } B \cong C \Rightarrow A \cong C,$$

then follows that $\mathcal{M}$ can be divided into **'similarity classes'**, which are disjoint to each other, and thereby each class consists of the totality of all n-level proportion chains, which are all similar to each other. (Because of reflexivity, by the way, all similarity classes are non-empty.) In mathematical language this means that

$$\mathcal{M}(A) = \{A' \in \mathcal{M} \mid A' \cong A\}$$

is the set of all n-level chains that are similar to a given chain A and which is then called the **similarity class for A.** Then we have the following structuring of the whole $\mathcal{M}$ as 'disjoint union of all possible equivalence classes of its members' and this can be expressed as follows:

Similarity class structure of all n-level proportion chains $\mathcal{M}$
  (a) $\mathcal{M} = \cup \{\mathcal{M}(A) \mid A \in \mathcal{M}\}$, and for all $A \in \mathcal{M}$ applies $\mathcal{M}(A) \neq \varnothing$.
  (b) $A \cong B \Leftrightarrow \mathcal{M}(A) = \mathcal{M}(B)$.
  (c) $A \not\cong B \Leftrightarrow \mathcal{M}(A) \cap \mathcal{M}(B) = \varnothing$.
For example, the last statement says that all chains similar to A are different from all chains similar to B if A and B are not themselves similar.

2. **The similarity criteria**
    Two n-level chains are similar if and only if one of the following criteria is met
        1. the step criterion,
        2. the magnitude criterion,
        3. or – in the case of commensurable proportion chains – the value criterion
    but all three are by the way equivalent and are described in Definition (2.2) of Sect. 2.1 and the sentence therein.

3. **Proportion-chains and their reciprocals**
    (a) **Invariance under Similarity:** For reversible proportion chains A holds

$$\mathcal{M}(A^{\text{rez}}) = (\mathcal{M}(A))^{\text{rez}} \text{ and } (\mathcal{M}(A^{\text{rez}}))^{\text{rez}} = \mathcal{M}(A).$$

    (b) **Symmetry:** The property of a proportion chain symmetry reads as follows:

$$A \text{ is symmetrical} \Leftrightarrow \text{all } A' \in \mathcal{M}(A) \text{ are symmetrical}$$
$$\Leftrightarrow \mathcal{M}(A^{\text{rez}}) \cap \mathcal{M}(A) \neq \varnothing \Leftrightarrow \mathcal{M}(A^{\text{rez}}) = \mathcal{M}(A).$$

(continued)

Theorem 2.6 (continued)

Thereby it can be said that as soon as there is a common chain of proportions in the similarity classes of A and of one of its reciprocals, it is already clear that A must be symmetrical. Furthermore it has to be said, that conversely, in the case of symmetry of A, both similarity classes are identical.

4. **Universality of Adjunction under Similarity**

   If for A and B an adjunction $C = A \oplus B$ is constructable so also for the complete similarity classes of these two proportion chains. Thus the global equation holds universally

$$\mathcal{M}(C) = \mathcal{M}(A) \oplus \mathcal{M}(B),$$

which states in detail:
1. For every chain A' from $\mathcal{M}(A)$ and for every chain B' from $\mathcal{M}(B)$ an composition $C' = A' \oplus B'$ is possible in the first place.
2. Each chain C' from $\mathcal{M}(C)$ is an composition of any two chains A' from $\mathcal{M}(A)$ and B' from $\mathcal{M}(B)$.

5. **Composition and Reciprocal**

   Let A be an n-stage proportion chain and B *be* an m-stage proportion chain, then there are equivalent
   (a) There exists an composition $C = A \oplus B$, and C is reversible.
   (b) There exists an composition $\bar{C} \cong B^{\mathrm{rez}} \oplus A^{\mathrm{rez}}$, and $\overline{C}$ is reversible.

   Conclusion: If statement (a) and consequently also statement (b) are true, then A and B are also themselves reversible, and the $(n + m)$-step chain C is a reciprocal of $\overline{C}$ and vice versa. As a result the important formula

$$(A \oplus B)^{\mathrm{rez}} \cong B^{\mathrm{rez}} \oplus A^{\mathrm{rez}}$$

holds and since all this is invariant under similarity the general set formula applies

$$(\mathcal{M}(A) \oplus \mathcal{M}(B))^{\mathrm{rez}} = (\mathcal{M}(B))^{\mathrm{rez}} \oplus (\mathcal{M}(A))^{\mathrm{rez}} = \mathcal{M}(B^{\mathrm{rez}}) \oplus \mathcal{M}(A^{\mathrm{rez}}),$$

provided that one of the properties (a) or (b) is true for only two representations: $A' \in \mathcal{M}(A)$ and $B' \in \mathcal{M}(B)$.

(continued)

Theorem 2.6 (continued)

6. **The Architecture of Symmetrical Proportion Chains**

For n-level proportion chains $A = a_0 : a_1 : \ldots : a_{n-1} : a_n$ the following architecture holds:

1. If n is even, that is, of the form $n = 2m$, then let

$$B = a_0 : a_1 : \ldots : a_{m-1} : a_m$$

is the front m-step subchain of A, and then

$$B^* = a_m : a_{m+1} : \ldots : a_{2m-1} : a_{2m}$$

is its mirrored proportion chain, and it is the posterior half of the chain A. Then the criterion holds:

$$A \text{ symmetrical} \Leftrightarrow B^* \cong B^{\text{rez}} \Leftrightarrow A = B \oplus B^{\text{rez}}.$$

2. If n is odd, that is, of the form $n = 2m + 1$, let

$$B = a_0 : a_1 : \ldots : a_{m-1} : a_m$$

again the front m-step subchains of A, and then the likewise m-step proportion chain is

$$B^* = a_{m+1} : a_{m+2} : \ldots : a_{2m} : a_{2m+1}$$

is their mirrored chain. If then $P = a_m : a_{m+1}$ is the middle step proportion of A, which connects B and B^* the criterion is valid:

$$A \text{ symmetrical} \Leftrightarrow B^* \cong B^{\text{rez}} \Leftrightarrow A = B \oplus P \oplus B^{\text{rez}}.$$

Proof Concerning (1): The three mentioned properties of reflexivity, symmetry and transitivity are indeed immediate transfers from the special case of ordinary proportions, as we have already encountered them in Theorem 1.1. Nevertheless, we want to justify this abstract object in detail in this new extended situation:

About symmetry: Because every proportion $a : b$ is similar to itself, this is also true for every proportion-chain A and therefore $A \cong A$, and from this it follows first of all that the family $\mathcal{M}(A)$ is not empty, because $A \in \mathcal{M}(A)$. It follows

$$\mathcal{M} = \{A \mid A \in \mathcal{M}\} = \cup_{A \in \mathcal{M}}\{A\} \subset \cup_{A \in \mathcal{M}} \mathcal{M}(A) \subset \cup_{A \in \mathcal{M}} \mathcal{M} \subset \mathcal{M},$$

so that the equality and thus the statement (a) holds. If a chain A is similar to chain B, then (by the transitive law for proportions) any chain similar to A is also similar to B and vice versa. Therefore, the similarity classes $\mathcal{M}(A)$ and $\mathcal{M}(B)$ are equal and statement (b) is shown. Finally, if two similarity classes $\mathcal{M}(A)$ and $\mathcal{M}(B)$ are not disjoint, then there is a chain C that is both in $\mathcal{M}(A)$ as well as in $\mathcal{M}(B)$. Then by definition $C \cong A$ and $C \cong B$, so again by the transitive law $A \cong B$ also follows. So non-similar proportion chains A and B cannot have similar chains in common, which also proves statement (c).

To (2): The two similarity-criteria (1) and (2) are equivalent by means of the cross-rule – we see the value-criterion like this: We set $a'_0 : a_0 = \lambda : 1$, then this is correct for all other proportions too. The cross-rule and a multiplication with λ then yields the proportion-relations $a'_k : 1 = \lambda a_k : 1$, which for numbers means nothing else than $a'_k = \lambda a_k$.

To (3): Here we have already shown all the statements; the set-theoretic formulation of the invariance property may take some getting used to: we shall therefore explain it briefly.

▶   The set $(\mathcal{M}(A))^{\text{rez}}$ is the set of all chains K that are reciprocal to all proportion chains A' that lie in $\mathcal{M}(A)$ – that is, to all those which are always similar to A. Since such a chain A' has step proportions similar to the corresponding step proportions of A, any chain K that is reciprocal to A' is also reciprocal to A and therefore similar to any chain reciprocal to A. Therefore K is also an element of the totality $\mathcal{M}(A^{\text{rez}})$ of all chains similar to A^{rez}, where A^{rez} is itself again any reciprocal to A.

All arguments are now equivalences and the first set equality

$$(\mathcal{M}(A))^{\text{rez}} = \mathcal{M}(A^{\text{rez}})$$

is shown. Due to the symmetry of similarity $(A^{\text{rez}})^{\text{rez}} \cong A$, which states that a reciprocal to the reciprocal of a chain A must again be similar to A, it also immediately follows that there is an equality of

$$(\mathcal{M}(A^{\text{rez}}))^{\text{rez}} = \mathcal{M}(A),$$

and point (3) is complete.

To (4): We already find this block of statements in Theorem 2.1.

To (5): In the first statement (a), one argues as follows:

If an composition $C = A \oplus B$ exists, this means in detail for the chains

$$A = a_0 : a_1 : \ldots : a_n \quad \text{and} \quad B = b_0 : b_1 : \ldots : b_m,$$

that the chain $C = c_0 : c_1 : \ldots c_n : c_{n+1} : \ldots : c_{n+m-1} : c_{n+m}$ the similar proportions

$$c_0 : c_1 \cong a_0 : a_1, \ldots, c_{n-1} : c_n \cong a_{n-1} : a_n$$
$$c_n : c_{n+1} \cong b_0 : b_1, \ldots, c_{n+m-1} : c_{n+m} \cong b_{m-1} : b_m$$

has. Now if C is invertible, there is an $(m + n)$-step chain $\overline{C}$ (its reciprocal),

$$\overline{C} = \overline{c}_0 : \overline{c}_1 : \ldots : \overline{c}_m : \overline{c}_{m+1} : \ldots : \overline{c}_{m+n-1} : \overline{c}_{m+n},$$

and by construction of a reciprocal, the front m-level subchain $\overline{c}_0 : \overline{c}_1 : \ldots : \overline{c}_m$ of $\overline{C}$ is the reciprocal of the back m-level subchain $c_n : \ldots : c_{n+m}$ of C. This, however, is similar to the m-level subchains B by definition of the composition $C = A \oplus B$. As a result the subchain $\overline{c}_0 : \overline{c}_1 : \ldots : \overline{c}_m$ is a reciprocal of B, and B is thus reversible. The exact same is true for the chain A: The rear n-level subchains $\overline{c}_m : \overline{c}_{m+1} : \ldots : \overline{c}_{m+n-1} : \overline{c}_{m+n}$ of $\overline{C}$ is reciprocal to the front n-level subchain $c_0 : c_1 : \ldots : c_n$ of C, which is again similar to the n-level chain A *after* the construction of an composition $C = A \oplus B$. Therefore A is also reciprocal.

Thus, from the existence of the composition and its reversibility, it follows that both chains A and B are reversible and that the $(m + n)$-step chain $\overline{C}$ is obviously the composition of the two chain reversals $\overline{B}$ and $\overline{A}$, where $\overline{A}$ is attached to $\overline{B}$. Thus '(a) $\Rightarrow$ (b)' is shown.

The reverse direction is completely analogous due to the symmetry $(A^{\text{rez}})^{\text{rez}} \cong A$ (because you can replace A by A^{rez} and B by B^{rez} and then apply to it what has already been shown).

To (6): Let $n = 2m$ be an even number of steps. Then we write the chain A in the form

$$A = a_0 : a_1 : \cdots : a_n = a_0 : a_1 : \cdots : a_m : a_{m+1} : a_{m+2} : \cdots : a_{n=2m},$$

and with the given decompositions

$$B = a_0 : a_1 : \ldots : a_m \quad \text{and} \quad C = a_m : a_1 : \ldots : a_n,$$

which are obviously directly connectable, the formula $A = B \oplus C$ is valid. But now this partial chain C is exactly the chain of the mirrored magnitudes of B formed in mirrored arrangement. In this case the mirrorings

$$a_0^* = a_n, a_1^* = a_{n-1}, \ldots, a_m^* = a_m,$$

are valid, so that the m-step partial proportion chain

$$B^* = a_m^* : a_{m-1}^* : \ldots : a_0^* = a_m : a_{m+1} : a_{m+2} : \ldots : a_{2m} = C$$

are obtained. Now, to show that B^* is reciprocal to B exactly when A is symmetric, we could use the similarity condition of diametric step proportions, but:

An abstract argument also wants to come into its own, and for this we use the conjunction of adjuncts and reciprocals from Part 5 of this theorem:

$$A = A^{\mathrm{rez}} \Leftrightarrow B \oplus B^* = (B \oplus B^*)^{\mathrm{rez}} \cong (B^*)^{\mathrm{rez}} \oplus B^{\mathrm{rez}}.$$

Therefore $(B^*)^{\mathrm{rez}} = B$ as well as $B^* = B^{\mathrm{rez}}$ is valid, but the latter is again equivalent to the former.

In the case of an odd number of steps $n = 2m + 1$ the choice of the m-step partial proportion chains then yields

$$B = a_0 : a_1 : \ldots : a_m \quad \text{and} \quad C = a_{m+1} : a_{m+2} : \ldots : a_{n=2m+1}$$

together with the proportion connecting them $P = a_m : a_{m+1}$ the composition of A in the sum $A = B \oplus P \oplus C$. Again, the chains B and C are mirrored, $C = B^*$. Combined with the trivial property of ordinary proportions to be reciprocal to themselves we find likewise with the preceding abstract argument

$$A \cong A^{\mathrm{rez}} \Leftrightarrow B \oplus P \oplus B^* \cong (B \oplus P \oplus B^*)^{\mathrm{rez}}$$
$$\Leftrightarrow (B^*)^{\mathrm{rez}} \oplus P^{\mathrm{rez}} \oplus B^{\mathrm{rez}} \cong (B^*)^{\mathrm{rez}} \oplus P \oplus B^{\mathrm{rez}}$$
$$\Leftrightarrow B^{\mathrm{rez}} \cong B^*.$$

Because of these important formulas, everything necessary is shown and the theorem is proved.

Among the undoubtedly numerous applications of this compendium of propositions, we pick out for once the architecture of symmetrical chains (that is, the statement (6)) and show in a concluding example how this theory of formulas feels in musical practice.

Example 2.7 Major and Minor Architecture in Chordal Music

Let's build a symmetrical chord structure from any starting note – let's say g_0. To do this, we add the major chord $g_0 - h_0 - d_1$ upwards to the tonic (note g_0) according to the proportion $D = 4 : 5 : 6$, and the minor chord $g_0 - es_0 - c_0$ downwards according to the proportion $M = 15 : 12 : 10$. Then, in upward succession, a minor triad is created in which a major triad is placed on the fifth (dominant). Then apparently

$$M^{inv} = D^{rez},$$

and the mathematical result is a proportion chain

$$A = M^{inv} \oplus D = D^{rez} \oplus D \cong 40 : 48 : 60 : 75 : 90,$$

which has the ambitus of the total proportions $4 : 9$, that is a Pythagorean ninth. The chain A is symmetrical, since it was constructed as such. Nevertheless, a brief check of the data confirms this. Likewise, we recognize that the magnitude corresponding to the tonic (that is in this case, the number 60) is the 'geometric center' of the chain: we have the symmetries

$$40 : 60 \cong 60 : 90 \text{ and } 48 : 60 \cong 60 : 75,$$

which, in anticipation of subsequent explanations, means nothing other than that the number 60 is the geometric mean of the two pairs of data $(40, 90)$ and $(48, 75)$. The musical result is a seventh-ninth minor chord in the step structure

minor third $(5 : 6) \oplus$ *major third* $(4 : 5) \oplus$ *major third* $(4 : 5) \oplus$ *minor third* $(5 : 6)$,

and here we have, for example, the tone sequence

$$c_0 - es_0 - g_0 - h_0 - d_1.$$

Of course, we could also make a construction similar to this by exchanging major and minor; then we would get the symmetrical proportion chain

$$B = D \oplus M = M^{rez} \oplus M \cong 20 : 25 : 30 : 36 : 45,$$

and here the magnitude – the initial tonic – is also the center of symmetry. The musical result is a 'dominant-seventh-ninth chord', which is here a diminished (major) seventh chord with ajoutated ninth, of the tone sequence

$$c_0 - e_0 - g_0 - b_0 - d_1$$

So especially the chord B is part of uncounted harmonic movements. ◄

An Echo

Done! There is no doubt that we have invested a great deal in this chapter in letting **mathematics play** the role of guardian of the clarity of concepts and of the logic of their interrelationships: **Plausibility crystalises** less strict as her sister and has had to exercise a little patience. Nevertheless, both have their right. And this is exactly what we ask of our readership: Some cannot get enough in the desire to understand things in their most abstract forms possible – others find their joy in analogy and its quasi-interpretations. It's very nice when both make an effort to understand the other side.

Medieties - Mean Value Proportion Chains

3

*A consonance is the measure from a low note to a high note, or
from a high note to a low note, which produces a melody.
(Cassiodorus, from [5], p. 127)*

When we hear the word 'mean' (our current term for mediocrity) we probably think of
certain averages of various data sets and a fearful premonition of possible totally boring
statistical interpolation tables and distributions possibly spreads first. Far from it!

▶ **Important**

*Rather, the structure of ancient music theory consists in no small part of an extremely
artful incorporation of what are known as the 'Babylonian medieties' or the
'medieties of Archytas' – namely, the*

- *arithmetic mean,*
- *harmonic means,*
- *geometrical means.*

*and some others – into the system of intervals, chords, tetrachordics and their scales built
from them.*

Of course, mean values also find a thousand applications outside music-theory: Besides
statistics, these are arithmetic and geometry (one only has to think of the well) known ray
theorems as well as the Pythagorean theorem group, which in turn forms an essential basic
concept of the entire elementary geometry. Already in the simplest (i.e. in the Pythagorean)
canon of the *sacred* numbers

© Springer-Verlag GmbH Germany, part of Springer Nature 2024
K. Schüffler, *Proportions and Their Music*,
https://doi.org/10.1007/978-3-662-65336-4_3

$$6 - 8 - 9 - 12 \quad \text{(Pythagorean canon)}$$

the two inner numbers represent the harmonic mean (the 8) and the arithmetic mean (the 9) of the octave numbers 6 and 12. At least in the ancient considerations left an enormous room for speculative, mystical as well as number-symbolic interpretations. And that was only the beginning. If, for example, we bring the pure major third into play (as is well known, it has the (pitch) proportion 4 : 5) this is expressed in the canon by adding the number 10 there. For in the now 'diatonic' canon

$$6 - 8 - 9 - 10 - 12 \quad \text{(diatonic canon)}$$

the proportion, which is similar to 4 : 5 arises with 8 : 10 and also the minor pure third (and some other intervals more) has found its place there with $10 : 12 \cong 5 : 6$. We now observe the following remarkable juxtapositions:

- The harmonic mean – i.e. the number 8 – divides the span $6 - 12$ into the two Sects. $6 - 8$ and $8 - 12$, which are themselves in the ratio $2 : 4 \cong 6 : 12$.
- The number 10, on the other hand, divides the span $6 - 12$ into the two Sects. $6 - 10$ and $10 - 12$, which are now in 'contra-ratio' $4 : 2 \cong 12 : 6$.

This number 10 is also a 'mean' – namely, the so-called 'contra-harmonic mean' of the numbers 6 and 12. The division ratio of the sections is the reverse of that of the harmonic mean and its position is also positioned in the opposite direction if we take the arithmetic mean as the new mean.

This simple experiment shows us in which direction the discussion with the medieties can probably lead and we will discover a whole range of surprising connections.

Usually, mean values are defined by formulas. However, verbal descriptions are not uncommon either – quite the contrary. In addition to the representations from the theory of proportions, antiquity almost without exception only knew the form of not infrequently vague paraphrases. Known descriptions are in any case in the case of two positive dates (a and b):

- The arithmetic mean x_{arith} is half of their sum,
- the harmonic mean y_{harm} is the reciprocal of the arithmetic mean of the reciprocals – but also the side which with the arithmetic mean gives a rectangle equal to the rectangle with sides a and b,
- the geometric mean z_{geom} is the side of a square equal to the rectangle with sides a and b.

We will describe the concepts of these and other averages not by means of statistical formulas, but largely by means of the rules and methods of proportion theory. And although these descriptions refer mainly to (numerical) proportions, a connection to more general magnitude proportions is possible throughout. Symbols such as '+' and '−' and some others would have to be interpreted accordingly. However, in music-theory it is 'frequency ratios' that are important: In this respect we encounter here rather only the situation in which the 'horoi' a, b, etc. are positive numbers. And as long as we move in ancient music-theory, these numbers are at best rational (fractions), if not only of the type of Perissos or Artios numbers: n and $1/n$. Whereby, due to the similarity ratios, even the restriction to integers (natural) would suffice.

Starting from the calculus of proportions, we first of all introduce

1. the general historical mean value proportions,
2. the Babylonian medietary trinity (geometric-arithmetic-harmonic),
3. the ten historical mean values - called: medieties.

and then develop the interrelationships which are essential for our purposes. In the case that a magnitude (x) is a mediety of a and b of a certain type, then the

Mean Value Proportion Chain $a : x : b$

in the focus of our interest, and we search for relations between such mean value proportion-chains for different medietary types: What are the common properties and what interrelationships can be obtained? Here, a highly fruitful connection between *mean values as magnitudes and their partition parameters* emerges.

And these interrelationships are explained by a functional interplay between the piece-to-rest proportion function

$$y = f(x) = (x - a)/(b - x),$$

which assigns its division parameter y to a magnitude ('mediateness x'), and its inverse, which assigns its mean values to the division parameters.

From a historical point of view, the theorems of **Iamblichos** and **Nicomachus of Gerasa** dominate the connection between Babylonian averages and musical-proportions. We shall see that these theorems are special cases of more general symmetries found in medietic proportion-chains. The two theorems of Iamblichos and Nicomachus, however, are at the heart of the ancient theory of musical proportions and their intervals, and they represent in content what was ultimately understood by the **'harmonia perfecta maxima'** **in** the most original and simplest case. We report on this in Sect. 3.2 in a first theorem, which has the Pythagorean canon as its model.

▶   However, using our results concerning the symmetry properties of medietic proportion-chains, we can embed this traditional harmonia perfecta maxima in a ***universal concept of a music theory*** that draws its richness precisely from the mutual relationships of proportion-chains, reciprocals, mean values and center symmetries that we have worked out. And geometry is also on board here, namely in the form of the *hyperbola of Archytas,* which will be a helpful companion for us later and especially in the network of further mean value and interval constructions.

The central objects of this chapter are thus, besides the presentation of ancient medieties, first and foremost the symmetries of Babylonian chains of proportions, and here a separate Sect. 3.5 is devoted to geometrical means. In it we come to perhaps the most modern form of the Pythagorean canon, the **'Harmonia perfecta maxima abstracta'**. It combines the magic of the ancient numerical series

$$6 - 8 - 9 - 12$$

with the magic of a magnitude series controlled by the geometric mean

$$a - x_1 - x_2 - b,$$

in which the inner magnitudes obey the laws of symmetry of the geometric mediety. Then follows in the last Sect. 3.6 of this chapter the description of the diatonic canon

$$6 - 7,2 - 8 - 9 - 10 - 12$$

in its most general classical mean form, **a musical résumé of the theory of proportions,** so to speak, and concludes the historical apparatus of ancient interval theory with a comprehensive basic understanding of Pythagorean diatonic music. At the same time, however, this prepares a new path –towards the **infinite musical medietic proportion-chains of** the inventive theorists, to which we will then turn in the following and final mathematical chapter (Chap. 4).

3.1   Mean Value Proportions and Their Analysis

The variety of historical meditations is quite considerable. But, thank God, one can recognize certain basic types in many of them, since they are either derived from the classical three meditations (the Babylonian meditations) or stand in some analogy to them.

It quickly becomes apparent that the **approach via the chains of proportions does** not only have its historical justification, but that this approach also represents a generalizing concept which allows for completely different ways of looking at things than the one which merely gives the mean values for given data as **mere formulas.**

This description of the medieties in the language of the theory of proportions can now make use of exactly three forms, viz.

- 'Piece to the rest' proportion,
- 'Piece to the whole' proportion,
- 'Piece to the piece' Proportion.

And in the literature one encounters (besides verbal and sometimes ornamental descriptions) exactly these three forms. It is obvious that only a uniformity for comparative considerations promises success. Although we already have described the basic interrelationships of these three forms of proportions in Theorem 1.3 from Sect. 1.3, we nevertheless want to mention this again in the case of the proportions of the medieties. This is done in the following theorem:

Proposition 3.1 (The Ancient Descriptions of Mean Value Proportions)

1. **Basic forms:** Let $a : x : b$ be a proportion chain with $a < x < b$, and let α, β be further magnitudes for which there is a ratio $\alpha : \beta$. We will call them **division parameters**. Then the following three forms are equivalent to each other:

   (A) The 'piece to the rest' form:

   $$(x-a) : (b-x) \cong \alpha : \beta \text{ or equivalently } (b-x) : (x-a) \cong \beta : \alpha.$$

   The 'piece to the whole' form:

   $$(x-a) : (b-a) \cong \alpha : (\alpha+\beta) \text{ or equivalently } (b-x) : (b-a) \cong \beta : (\alpha+\beta).$$

   (B) The 'piece to piece' form:

   $$(x-a) : \alpha \cong (b-x) : \beta \cong (b-a) : (\alpha+\beta).$$

   The forms (A) thus describe the relation of the two subsections $(x - a)$ (the 'piece') and $(b - x)$ (the 'rest'), created by the 'mean' x and complementary to each other, to each other. The forms (B) describe the relation of both sections to the 'whole' $(b - a)$ – where from one follows the other. Finally, forms (C) give the relations of all three parts to the division parameters α and β, where $\alpha + \beta$ is interpreted as the 'whole' with α as the 'piece' and β as its 'rest' in the whole. For illustrations, see also Fig. 3.1.

2. **Monotonicity of Mean Values, Position Criterion:** If we have two mean value proportions for the same data a and b, which are, for example, in the piece-to-rest form (A),

   $$(x-a) : (b-x) \cong \alpha : \beta \text{ and } (\widetilde{x}-a) : (b-\widetilde{x}) \cong \widetilde{\alpha} : \widetilde{\beta},$$

(continued)

Proposition 3.1 (continued)
then the following applies: The magnitude ratios of the division parameters are directly transferred to the corresponding mean values and this thereby causes an ordering of the magnitudes – namely: The more exact equivalence relation applies:

$$\frac{\alpha}{\beta} = \lambda < \tilde{\lambda} = \frac{\tilde{\alpha}}{\tilde{\beta}} \Leftrightarrow a < x < \tilde{x} < b.$$

Conclusion: A mean value (x) is **smaller** than another ($\tilde{x}$), exactly if the frequency measure $\frac{\beta}{\alpha}$ of its division parameter proportion $\alpha : \beta$ is **larger** than that of the other mean value.

Proof to (1): Indeed, we can obtain the equivalence of the given representations from the rules of the theory of proportions. We get this result by applying the 'piece' formulas there: To do this, we simply put in the piece formulas of Theorem 1.3

$$X = (x - a) \text{ and } Y = (b - x)$$

optionally for 'piece' and 'rest', then the 'whole' is equal to $X + Y = (b - a)$. Similarly, the two parameters α and β are 'piece' and 'rest' of the 'whole*' $\alpha + \beta$, and the first part of the theorem follows.

We simply take the part (2) from the view and have the case (A) present, with which the two partial sections $(x - a)$ (the 'piece') and $(b - x)$ (the 'rest') stand in proportion to each other, as the sketch of the Fig. 3.1 shows it to us before eyes. Since both sections add up to the given total size $(b - a)$ (the 'whole'), we can then also very easily read off the position of the mean values comparatively. A simple observation is sufficient for the time being:

Geometrically Motivated Consideration
Let us imagine the distance from *a* to *b* on the number line divided by an intermediate point *x* into the two sections $(x - a)$ and $(b - x)$, as shown in Fig. 3.1. If *x* now moves upwards in the direction of *b*, $(x - a)$ becomes larger – but $(b - x)$ becomes smaller at the same time. Both causes that the quotient (the division parameter)

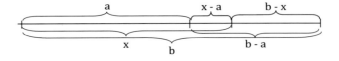

Fig. 3.1 Distance model of the proportion function

$$(x-a)/(b-x)=\lambda$$

grows, since its numerator becomes larger and its denominator smaller. Accordingly, the reverse is true if the intermediate point x moves towards a: Here the section $(x - a)$ becomes smaller and $(b - x)$ simultaneously gets larger. As a result the fraction consequently gets smaller. And for the reciprocal of λ (the frequency measure of the division parameter proportion) the reverse growth behavior logically applies.

Although we have used the interpretation of proportions as numerical ratios and consequently fractional arithmetic for convenience, this observation could undoubtedly also be justified by the laws of proportions. Whereby it is to be noted that some of their argumentations would also grow out of the descriptively motivated laws of experience. Let us think, for example, of the smaller-larger relation, whose inclusion would only happen via the 'basic mathematical' definition of a complementary (positive) magnitude (z): $x < y \Leftrightarrow y = x + z$.

However, we benefit much more conveniently from the analysis of the **proportion function:**

$$y = f(x) = (x - a)/(b - x),$$

which is adapted to the 'piece to the rest' form and can lead us into 'modern' mathematics.

Remark   Here it is advisable to take the parameter $q := b/a$ on board as well. It appears in many formulas and almost all decision criteria are oriented (seen in the light of day) less on the magnitudes of the magnitudes a and b themselves, but rather and almost exclusively on their ratio (the 'proportion') to each other. This parameter q has (depending on the point of view and usage) many names: **Interval measure** or **frequency measure** as well as **frequency factor of** the musical interval $[a, b]$ or of the proportion $a : b$. And it can also be found in financial mathematics: There it is called the **'compounding factor'**.

Thus, apart from the ordinary analysis which clarifies the questions of continuity, monotonicity and reversibility, the following theorem establishes in particular two important connections between the variables and their function values, if we have in view the frequency measure analysis which is important for our concerns. Above all, however, the theorem provides, at the level of a fundamental and generally valid **'functional equation'**, the mathematical background of the symmetries of the mean values among themselves on the one hand and the symmetries of the piece-to-rest proportion parameters (the division parameters) corresponding to them on the other.

Here, the reputation of the geometric mean of being an *'undisputed ruler over all symmetries'* – in our estimative formulation – is unmistakably consolidated.

Theorem 3.1 (The Analysis of the Proportion Function and the Hyperbola of Archytas)

1. **The 'proportion function':** The real function

$$f(x) = (x-a)/(b-x)$$

assigns to each value x with $a < x < b$ the size ratio of the resulting mean value sections $(x-a)$ (the 'piece') versus $(b-x)$ (the 'remainder'). We will henceforth refer to this size ratio as the **division parameter** (for x). This function is continuous and arbitrarily differentiable in the variable domain $a \leq x < b$ and has a simple zero in $x = a$. It is strictly monotonically increasing from 0 to $+\infty$ when the variable x is strictly monotonically increasing from a to b. At the point $x = b$ the function has a simple pole and its asymptotic is this:

$$f(x) \nearrow +\infty \text{ at } x \nearrow b.$$

2. **The 'mean value formula':** The equation

$$y = f(x) = (x-a)/(b-x)$$

has due to the course described in (1.) for each given value y – thus for each division parameter – with $0 \leq y < \infty$ exactly one solution x, which lies beyond that also in the interval $a \leq x < b$ and with the frequency measure $q := b/a$ of the considered proportion $a : b$ this solution reads as a **mean value formula**

$$x = g(y) = a + \frac{y}{1+y}(b-a) = \frac{a+yb}{1+y} = a\frac{1+yq}{1+y}.$$

The **mean value function** $g(y)$ determined in this way is continuous and differentiable any number of times on the positive number line $[0, \infty[$. Furthermore it is the inverse function of the proportion function $f(x)$: It thus assigns to a given proportion parameter $y > 0$ exactly that value x (**'mean value'**) between a and b *with* respect to which the two sections $(x-a)$ *and* $(b-x)$ just form the ratio y to each other. This is then also expressed by the 'inverse relations':

(continued)

Theorem 3.1 (continued)
$$f(x) = (x-a)/(b-x) = y \Leftrightarrow x = a + \frac{y}{1+y}(b-a),$$

and this applies to all $0 \leq y < \infty$ and to all $a \leq x < b$.

3. **The Monotonicity of the Mean Values:** As the inverse of a strictly monotonically growing function, the mean value function g is also strictly monotonically growing: If y moves monotonically from 0 to $+\infty$, the solution $x = g(y)$ of the proportion equation $f(x) = y$ also moves monotonically from a to b. All in all we have the inequalities

$$0 < y_1 < y_2 \Leftrightarrow a < x_1 := \frac{a + y_1 b}{1 + y_1} < x_2 := \frac{a + y_2 b}{1 + y_2} < b.$$

gain. This inequality represents the monotonicity of the **mean values.**

4. **The Frequency Measure Symmetry:** Between an average value x and its Proportion parameter $y = f(x)$ there is a remarkable symmetry with respect to bilateral frequency measure dependencies:

  Exactly when the relative size x/a depends only on the frequency measure $q = b/a$, the division parameter y also depends only on q, that is:

$$x/a = \psi(q) \Leftrightarrow y = f(x) = \varphi(q)$$

with suitable functions ψ and φ of the one variable q.
  The Functional Equation: For all $y > 0$ the 'functional equation'

$$g(y) * g(1/qy) = ab$$

is valid, whose importance lies mainly in the following relationship between mean values and their proportion parameters: Namely, they are equivalent:

(a) $x_1 * x_2 = ab$ – which means that both are mirrored : $x_2^* = x_1$,
(b) $f(x_1) * f(x_2) = a/b$.

Here we can write the condition b) also in the forms

$$f(x_2) = 1/qf(x_1) \text{ respectively } f(x_1) : \frac{1}{\sqrt{q}} \cong \frac{1}{\sqrt{q}} : f(x_2).$$

(continued)

Theorem 3.1 (continued)

Conclusion: Because the parameter $q^{-1/2} = \sqrt{a/b}$ is exactly the proportion parameter of the geometric mean $z_{geom} = \sqrt{ab}$, we can also express the above equivalence as follows:

$$x_1 : z_{geom} \cong z_{geom} : x_2 \Leftrightarrow f(x_1) : f(z_{geom}) \cong f(z_{geom}) : f(x_2),$$

which leads us to the remarkable symmetry:

Two mean values x_1 and x_2 lie symmetrically (mirrored) with respect to the geometric mean z_{geom}, if their associated division parameters are

$$y_1 = f(x_1) \text{ and } y_2 = f(x_2)$$

are symmetrical (mirrored) with respect to the proportion parameter $(q^{-1/2})$ of the geometric mean (z_{geom}).

5. **The Hyperbola of Archytas:** The graph of the real function

$$y = h(x) = \frac{ab}{x} \; (with \; x > 0)$$

is a hyperbola with the bisectors as symmetry axes.

Let us call it the **'hyperbola of Archytas'** and $y = h(x)$ the **Archytas Function** in connection with the theory of our Harmonia perfecta maxima. Now applies:

The hyperbola of Archytas is exactly the geometric locus of all variables x_1 and x_2, which are mirrored magnitudes with respect to the proportion $a : b$, thus:

$$x_1 * x_2 = ab \Leftrightarrow x_2^* = x_1$$
$$\Leftrightarrow \text{ the point } (x_1, x_2) \text{ lies on the hyperbola of Archytas.}$$

Explanation: The intersection of this hyperbola with the bisector $y = x$ is exactly the coordinate point (z_{geom}, z_{geom}), where

$$z_{geom} = \sqrt{ab} = a\sqrt{q} = aq^{1/2}$$

is the geometric mediety of a and b.

Before giving some further explanations of the proof, we show in which direction one or other of the consequences of these observations is moving:

(A) **Meaning of the frequency measure symmetry**

   Most of the equations of determination of the mean values can be expressed in the uniform form

$$(x-a)/(b-x)=y$$

where y *is* a 'given' division parameter for the complete interval of length $(b-a)$. It can be seen very often that this parameter can be written as an expression in the relative size b/a – but sometimes not. That one nevertheless wants to do this is due to the fact that a comparison of the position of various mean values is all the easier if this can be done for their division parameters. This is a consequence of the mean value monotonicity (3) of the theorem. In the case of a uniform functional description of these division parameters y as a function of the frequency measure q, this is usually much more convenient than some direct mean value inequalities, which can only be obtained with difficulty. We will demonstrate exactly this later – namely in the discussion of the contra-mediety-sequences in Sect. 4.2.

(B) **Meaning of the functional equation**

   Thus, if the geometric mean z_{geom} by definition the equation or proportion

$$ab=z_{geom}^2 \Leftrightarrow a:z_{geom} \cong z_{geom}:b$$

fulfilled, then we can write an equation $x_1 * x_2 = ab$, in the form as a proportion

$$x_1:z_{geom} \cong z_{geom}:x_2,$$

from which we see that the magnitudes x_1 and x_1 are both mirrored to the edges a and b and symmetrical to the geometric mean. And z_{geom} is therefore simultaneously also the geometric mean for x_1 and x_2. Later we will say that the proportion chains

$$x_1:z_{geom}:x_2 \text{ and } x_2:z_{geom}:x_1$$

are geometric. At the same time the proportion parameters are then mirrored values to the proportion parameter $\sqrt{a/b}$ of the geometric mean. Thus, if we specify an arbitrary proportion parameter $y>0$ for which we have the unique means x_1 and x_2 as solutions of the two mean value proportion equations

$$(x_1-a)/(b-x_1)=y \text{ and } (x_2-a)/(b-x_2)=1/qy=a/by$$

these satisfy the equation $x_1 * x_2 = ab$, which proves that the proportion chain

$$x_1 : z_{\text{geom}} : x_2 \text{ respectively } x_2 : z_{\text{geom}} : x_1$$

is a geometric proportionn chain. And conversely, the proportion parameters y_1 and y_2 of two magnitudes x_1 and x_2, whose product is $a * b$, have the unique relation

$$y_1 = a/by_2.$$

We can recognize the order of the magnitudes by the monotonicity: for the case that $0 < y < \sqrt{a/b}$ holds, we get the order $x_1 < z_{\text{geom}} < x_2$ and in the contrary case $y > \sqrt{a/b}$ we have the order $x_2 < z_{\text{geom}} < x_1$.

(C) **The geometry of the hyperbola of Archytas**

The graph of the well-known function ('Archytas function')

$$y = h(x) = ab/x,$$

which we consider only in the 1st positive quadrant $x > 0$, $y > 0$, is a true hyperbola, whose symmetry axes are both bisectors $x = y$ and $x = -y$. The hyperbola of Archytas can be looked up in the appendix. By the way, another familiar representation of the same hyperbola would be as follows:

$$(y + x)^2 - (y - x)^2 = 4ab.$$

It crosses the bisector $y = x$ perpendicularly at the point (z, z) with $z = \sqrt{ab}$. If, for example, (x_1, y_1) and (x_2, y_2) are points on this hyperbola with the arrangement

$$x_1 < x_2 < y_2 < y_1,$$

then first of all both points lie above the diagonal $y = x$ on the function graph of the Archytas function h, and thereby the point (x_1, y_1) lies also above the point (x_2, y_2). Thus, this is on the curve piece of the hyperbola between (x_1, y_1) and the point (z, z) of the geometric mean $z = z_{\text{geom}}$.

Remark: Finally, we remark that the naming of the familiar hyperbolic function h (x) is well founded in our context: **Archytas of Tarent** *(c. 435–350 BC) decisively consolidated the medietary structure around the mean value proportion $a : z \cong z : b$ of the geometric mean.*

(D) **The parameter representation of the hyperbola**

The vectorial mapping

$$G : \mathbb{R}_+ \rightarrow \mathbb{R}_+ \times \mathbb{R}_+, G(y) = (g(y), g(1/qy))$$

is a parametrization of the hyperbola of Archytas (namely between the points (a, b) and (b, a);). Because for all proportion parameters $y > 0$ the magnitude pair $(g(y), g(1/qy))$ lies on the hyperbola and if the parameter y runs from 0 to ∞, the image point $G(y)$ runs monotonically from (a, b) via the angle bisector point (z, z) – where the two points $g(y)$ and $g(1/qy)$ exchange roles – to the mirror point (b, a).

Proof of the Theorem
Helpful is first the transformation

$$f(x) = \frac{b-a}{b-x} - 1,$$

from which the graphical course is very quickly recognizable: The graph of the known function $f_1(x) = \frac{1}{-x}$ is shifted horizontally by the distance b to the right, then the function $f_2(x) = \frac{1}{b-x}$ is created. The subsequent multiplication by the constant positive factor $(b - a)$ does not change anything in the qualitative course of $f_2(x)$, only all function values are changed by this uniform factor $(b - a)$. The subsequent subtraction by (-1) simply means a vertical translation of the graph downwards by 1 unit. And the monotonicity for the proportion function is easiest to read off from the transformation – familiarity with the basic function 1/x once assumed.

Regarding statement (2): First of all the function f on the interval $[a, b]$ is continuous and strictly monotonically increasing and it has a zero in a including the obvious asymptotics

$$f(x) \nearrow +\infty \ \ at \ x \nearrow b.$$

As a result it follows now immediately from the intermediate value theorem for continuous functions that the equation

$$f(x) = (x - a)/(b - x) = y$$

has **exactly one (positive) solution** x for **each positive** y. By simply rearranging this equation according to the variable x, the mean value formula is obtained. The monotonicity of the proportion function is transferred 1 to 1 to its inverse function – i.e. to the mean value function $g(y)$, whereby also the mean value monotonicity (3) is obtained in full generality and without tedious calculation with inequalities.

We see statement (4) with the formulas from (1) and (2): Let $y = f(x) = \psi(q)$. Then it holds

$$f(x) = \frac{x-a}{b-x} = \frac{a(\psi(q) - 1)}{b - a\psi(q)} = \frac{\psi(q) - 1}{q - \psi(q)} =: \varphi(q),$$

because this last expression is obviously a function that depends only on q, and let this be denoted by $\varphi(q)$. Now if the reverse is true

$$y = f(x) = \varphi(q),$$

then the formula follows for $x/a = g(y)/a$

$$\frac{g(y)}{a} = \frac{g(\varphi(q))}{a} = \frac{1 + \varphi(q)q}{1 + \varphi(q)} =: \psi(q),$$

which appears to be a function $\psi(q)$ of the sole variable q.

Finally, we show the functional equation (5): We calculate

$$g(y) * g\left(\frac{1}{qy}\right) = a\frac{1 + yq}{1 + y} * a\frac{1 + (qy)^{-1}q}{1 + (qy)^{-1}} = a^2 \frac{1 + yq}{1 + y} * \frac{1 + (y)^{-1}}{1 + (qy)^{-1}}$$

$$= a^2 \frac{1 + yq}{1 + y} * \frac{(1 + y)(y)^{-1}}{(1 + qy)(qy)^{-1}} = a^2 * \frac{(y)^{-1}}{(qy)^{-1}} = a^2 q = ab,$$

and for this we needed only solid fractional arithmetic. The conclusion now results on the one hand from this monotonicity-consideration – concerning the position of the data x_1, x_2; by the way, the exceptional value $y = \sqrt{a/b}$ results in the trivial special case $x_1 = z = z_{\text{geom}} = x_2$ and then the claimed property follows from the functional equation, because yes with

$$x_1 := g(y) \quad \text{and} \quad x_2 := g\left(\frac{1}{qy}\right)$$

the equation $x_1 * x_2 = z^2$ is satisfied, which means nothing else than that the proportion $x_1 : z = z : x_2$ is geometric.

The statement point (6) arises from a simple consideration of Archytas function $h(x)$ and the evaluation of the preceding results and we leave the simple details to busy reworking. Thus the theorem is shown.

For illustration of the required functions, we refer to the appendix (figures), where the qualitative progressions of the proportion function $f(x)$, the mean function $g(y)$ and the hyperbola of Archytas $y = ab/x$ are presented. They are placed there because they may be useful for illustration at various points in our text.

In the following example, we cannot avoid using some of the mean values, the descriptions of which we will deal with in the next Sect. 3.2 and the section after next 3.3 respectively. However, for the moment a look at the enumeration of the Definition 3.2 there is sufficient for understanding.

Example 3.1 Functional Equation and Symmetry of the Proportion Parameters

Let x_1 and x_2 be respectively solutions of the following mean value proportions

1. $(x_1 - a) : (b - x_1) \cong a : b$ and $(x_2 - a) : (b - x_2) \cong 1 : 1$.

Then x_1 and x_2 are mirrored, $x_1 * x_2 = ab$, and it's $a < x_1 < x_2 < b$;
x_1 is the **Harmonic** and x_2 is the **Arithmetic** mean of a and b.
Why? Well, the proportion parameters are q^{-1} (for x_1) and q^0 (for x_2). They are
obviously in geometric proportion to $q^{-1/2}$, because the proportion of the geometric
mediety mediaty is valid

$$q^{-1} : q^{-1/2} \cong q^{-1/2} : 1,$$

and $q^{-1/2}$ is the geometric mean of q^{-1} and $q^0 = 1$.

2. $(x_1 - a) : (b - x_1) \cong a^2 : b^2$ and $(x_2 - a) : (b - x_2) \cong b : a$.

Then x_1 and x_2 are also mirrored, $x_1 * x_2 = ab$, and it's $a < x_1 < x_2 < b$;
x_1 is the **Contra-Arithmetic** and x_2 the **Contra-Harmonic** mean of a and b.
Why? Here the proportion parameters are q^{-2} (for x_1) and q^1 (for x_2); They too are
apparently in geometric proportion to $q^{-1/2}$, for we have the proportion

$$q^{-2} : q^{-1/2} \cong q^{-1/2} : q^1,$$

and here $q^{-1/2}$ is the geometric mean of q^{-2} and q^1.   ◀

The two examples show impressively that we have the crucial equation

$$x_1 * x_2 = ab$$

could show even **without calculation and knowledge of** the two mean values –only the
fundamentally much simpler consideration of their proportion parameters

$$\varphi(q) = f(x_1) \quad \text{and} \quad \psi(q) = f(x_2)$$

was sufficient for this purpose. In other words: The amusing guessing game is apparently

'Tell me your values and I'll tell you who you are'

the mathematical version of a sorcery from old fairy tales and fables.

3.2   The Trinity 'Geometric – Arithmetic – Harmonic' and Its Harmonia Perfecta Maxima

The significance of the chains of proportions in music theory derives above all from the play with the **proportions of the mean values** and their **chains.** Now, in antiquity there were indeed 'many' such chains of proportions –but from time immemorial, only three of them dominated events,

- the proportion chain of the geometric mediety,
- the proportion chain of the arithmetic mediety,
- the chain of proportions of the harmonic mediety.

These meditations are attributed to the Pythagorean Archytas (ca. 435–350 BC), among others, but other sources attest them a much older, Babylonian origin. We therefore coin for them the characterizing expression **'Babylonian Medieties'.** Archytas himself also mentions **Pythagoras** as a source.

From these famous medieties, the 'Harmonia perfecta maxima' arose in the Pythagorean octave canon and in its simplest and most fundamental form. In this section, we will analyze and prove this musical principle for **any given** canon. In doing so, we will also discover very interesting symmetries that exist between these chains of proportions and their reciprocals. Exactly this aspect we will explore more detailed in the continuing Sects. 3.6 and 4.5.

In the descriptions of the medieties, which (as already mentioned many times) usually only knew the verbal form, the imaginative use and meaningfulness of the formulation should (as in all these contexts) always be included! The following Definition 3.1 gives a sense of these ancient forms of description:

Definition 3.1 (The Babylonian Medieties: The Medieties of Archytas)

Let a, b, c and d be magnitudes – that is, magnitudes which can be compared by means of proportions, that is, which are in proportion to each other. Then one finds the following ancient-historical descriptions of mean values:

1. **Geometric Mediety:**
   'If in the four-membered proportion $a : b = d : c$ the middle members are equal, the proportion is called **continuous,** and $b = d$ is the **geometrical** mediety of the outer members (a and c).'

2. **Arithmetic Mediety:**
   'If the middle one (b) is as much more than the smaller one (a) as it is less than the larger one (c), it is the **arithmetical** mediety of the smaller and the larger (a and c).'

(continued)

Definition 3.1 (continued)

But this formulation is also well known:

'The middle one – taken twice – is as much as the smaller one and the larger one put together.'

3. **Harmonic Mediety:**

*'If the rest of the smaller (a) in the middle (b) relates to the rest of the middle (b) in the larger (c) as the smaller relates to the larger, the middle is the **harmonic** mediety of the smaller and larger.'*

It is then said that the proportion chain $a : b : c$ is **geometric** or **arithmetic** or **harmonic**.

Conclusion: We can easily transfer these descriptions into the language of proportions: Accordingly, the following definitions apply to an (ascending) chain of proportions $a : x : b$.

1. The magnitude x is the **Geometric Mediety** of a and b and we denote this as $x = z_{geom}$ - in the mathematical language:

$$x = z_{geom} \Leftrightarrow a : x \cong x : b.$$

2. The magnitude x is the **Arithmetic Mediety** of a and b and we denote this as $x = x_{arith}$ in the mathematical language

$$x = x_{arith} \Leftrightarrow (x - a) : (b - x) \cong 1 : 1.$$

3. The magnitude x is the **Harmonic Mediety** of a and b and we denote this as $x = y_{harm}$ - in the mathematical language

$$x = y_{harm} \Leftrightarrow (x - a) : (b - x) \cong a : b.$$

The chain of proportions $a : x : b$ is then also called **a geometrical** or **arithmetical** or **harmonic** chain of proportions or medieties. And together we call them **'Babylonian medietary chains'**.

In fact, much of this is already contained in the simple **Pythagorean canon**

$$6 - 8 - 9 - 12$$

can be read. This consists precisely of the harmonic (8) and the arithmetic (9) mean of the octave numbers 6 and 12. The following is astonishing is above all:

The entire Pythagorean theory of music is developed from this canon alone. At the center of this are the equations of proportion

$$6 : 8 \cong 9 : 12 \ \text{ or } \ 6 : 9 \cong 8 : 12.$$

These Pythagorean proportion equations, which became known as **Harmonia perfecta maxima,** are special cases of the theorems of **Iamblichos** and **Nicomachus,** which we will now describe (but in a generalized form). They, in turn, lead directly to the **authentic-plagal decomposition of musical intervals,** concepts that we also know in connection with Gregorian scales (modes).

Theorem 3.2 (The Harmonia Perfecta Maxima of the Babylonian Canon)
Let a and b be two magnitudes. If

$$x_{\text{arith}} \text{ denotes the arithmetic mediety for } a \text{ and } b,$$
$$y_{\text{harm}} \text{ denotes the harmonic mediety for } a \text{ and } b,$$

so we consider the 3-step proportion chain formed with these two medieties

$$P_{\text{mus}} = a : y_{\text{harm}} : x_{\text{arith}} : b,$$

which Iamblichos already called (in the case of the octave canon $6 : 8 : 9 : 12$) **'the musical chain of proportions'.** Other names are **Babylonian** (or also Pythagorean) **proportion chain** (respectively **Babylonian canon**). In this canon, the following symmetry prevails:

(A) **Theorem of Iamblichos of Chalcis (245–325 AD):**
The following similarities of proportions (equivalent thanks to the cross-rule) apply

$$a : y_{\text{harm}} \cong x_{\text{arith}} : b,$$
$$a : x_{\text{arith}} \cong y_{\text{harm}} : b,$$

which we can also formulate in the following way: The arithmetic and the harmonic medieties of a proportion are always mirrored magnitudes to each other,

$$x^*_{\text{arith}} = y_{\text{harm}} \ \text{ and } \ y^*_{\text{harm}} = x_{\text{arith}}.$$

Conclusion: The proportion chain P_{mus} is symmetrical, i.e. reciprocal to itself:

(continued)

Theorem 3.2 (continued)

$$P_{\text{mus}} \cong P_{\text{mus}}^{\text{rez}}.$$

(B) **Theorem of Nicomachus of Gerasa (c. 60–120 AD):**

Moreover, if the magnitude z_{geom} is the geometric mediety belonging to a and b, so that the ascending magnitude sequence

$$a < y_{\text{harm}} < z_{\text{geom}} < x_{\text{arith}} < b$$

is created, then the proportions also apply

$$y_{\text{harm}} : z_{\text{geom}} \cong z_{\text{geom}} : x_{\text{arith}}.$$

Simultaneously, this means that the geometric mean z_{geom} is also the geometric mean of the medieties y_{harm} and x_{arith}, and the chain of proportions formed by these three Babylonian medieties

$$y_{\text{harm}} : z_{\text{geom}} : x_{\text{arith}}$$

is then itself a geometrical chain of proportions, because this is the exact definition of such a chain.

(C) **Connection of both theorems:** In case of the existence of the geometric mean (as an 'writable' magnitude) both theorems are also equivalent, from which we obtain the authentic-plagalic decomposition of musical intervals: For a given musical intervall a : b construct the musical Pythagorean proportion chain described earlier P_{mus}. If then

$$A_{\text{arith}} \cong a : x_{\text{arith}} \cong y_{\text{harm}} : b$$
$$A_{\text{harm}} \cong a : y_{\text{harm}} \cong x_{\text{arith}} : b$$

are the proportions formed to these two medieties, then the two chains of proportions

$$A_{\text{auth}} \cong a : x_{\text{arith}} : b \cong A_{\text{arith}} \oplus A_{\text{harm}}$$
$$A_{\text{plag}} \cong a : y_{\text{harm}} : b \cong A_{\text{harm}} \oplus A_{\text{arith}}$$

are reciprocal to each other. They have an inverted architectural structure.

(continued)

Theorem 3.2 (continued)

Equivalent to this it can be said that the outer overall proportion $A = a : b$ can be divided by means of the medieties x_{arith} and y_{harm} respectively into 2-step chord-chains A_{auth} and A_{plag}, whose step proportions are similar when interchanged, which is called **Authentic Division**, 'major model' or **Plagal division**, 'minor model'– or decomposition – of the musical interval $a : b$ respectively.

Remark

In antiquity, these general relations represented 'theorems' (at least for the evident number case of the octave) that is, for the Pythagorean canon.

We now want to prove this exclusively by means of the laws of proportion (the rules from Theorem 1.3) in order to follow in the footsteps of the ancient teachers. With the well known mean value formulas

$$2x_{arith} = (a + b) \text{ and } y_{harm} = 2ab/(a + b),$$

which we will shortly derive among many other formulas, the proof of the proportions and their harmonia would, however, be done virtually by 'looking'.

Derivation

First we show the part (A) and this is done in single steps:

1. For the harmonic mediety y_{harm} the relationship applies:

$$a : y_{harm} \cong (a + b) : (b + b).$$

Because the definition

$$(y_{harm} - a) : (b - y_{harm}) \cong a : b$$

passes with the sum-rule, then with the cross-rule and then with the inverse-rule equivalently into the forms

$$y_{harm} : (b + b - y_{harm}) \cong a : b \Leftrightarrow y_{harm} : a \cong (b + b - y_{harm}) : b$$
$$\Leftrightarrow a : y_{harm} \cong b : (b + b - y_{harm}),$$

and applying the sum-rule again gives the required equation.

2. For the arithmetic mediety x_{arith} the relation holds:

$$x_{\text{arith}} : b \cong (a+b) : (b+b).$$

Because from the definition arise with the cross-rule first the similarities

$$(x_{\text{arith}} - a) : a \cong (b - x_{\text{arith}}) : a \Leftrightarrow (x_{\text{arith}} - a) : (b - x_{\text{arith}}) \cong a : a,$$

and the sum-rule and the subsequent inversion then yield the forms

$$x_{\text{arith}} : (a+b-x_{\text{arith}}) \cong a : a \Leftrightarrow (a+b-x_{\text{arith}}) : x_{\text{arith}} \cong a : a \cong b : b.$$

In fact, $a : a \cong b : b$, because namely $a : a \cong b : b \Leftrightarrow a : b \cong a : b$ is due to the cross-rule for repetition. This last trick ('replace $a : a$ by the proportion $b : b$') leads with the cross-rule to the equation

$$(a+b-x_{\text{arith}}) : b \cong x_{\text{arith}} : b,$$

and again with the sum-rule we arrive at the assertion

$$(a+b) : (b+b) \cong x_{\text{arith}} : b.$$

3. Now the proportion equation of Nicomachus follows because of the symmetry and transitivity-rules of the laws of proportion:

$$a : y_{\text{harm}} \cong (a+b) : (b+b) \cong x_{\text{arith}} : b \Rightarrow a : y_{\text{harm}} \cong x_{\text{arith}} : b.$$

Consequently, it follows that the musical chain of proportions B_{mus} is symmetrical and Iamblichos' theorem is proved along with its equivalent forms.

For further consideration in part (B), the statement block (4) now serves:

4. From the defining equation $z_{\text{geom}} : a \cong b : z_{\text{geom}}$ for the geometric mean as well as with the help of the proportion fusion the similarities

$$a : y_{\text{harm}} \cong x_{\text{arith}} : b$$
$$\Leftrightarrow (a : y_{\text{harm}}) \odot (z_{\text{geom}} : a) \cong (x_{\text{arith}} : b) \odot (z_{\text{geom}} : a)$$
$$\Leftrightarrow z_{\text{geom}} : y_{\text{harm}} \cong (x_{\text{arith}} : b) \odot (b : z_{\text{geom}}) \cong x_{\text{arith}} : z_{\text{geom}}$$

will follow with part (A). Therefore

$$z_{\text{geom}} : y_{\text{harm}} \cong x_{\text{arith}} : z_{\text{geom}} \text{ or } y_{\text{harm}} : z_{\text{geom}} \cong z_{\text{geom}} : x_{\text{arith}},$$

and the chain $y_{\text{harm}} : z_{\text{geom}} : x_{\text{arith}}$ is geometric.

The addition (C) is seen like this: If the similarity of proportions

$$y_{harm} : z_{geom} \cong z_{geom} : x_{arith},$$

yields, it follows that the equation

$$y_{harm} * x_{arith} = z_{geom}^2 = ab$$

holds. Therefore the magnitude pair (x_{arith}, y_{harm}) lies on the hyperbola of Archytas, and simultaneously the following equation of Iamblichos is valid:

$$a : y_{harm} \cong x_{arith} : b,$$

which proves our theorem. The statements about the authentic and plagalic decomposition of musical intervals and proportions, respectively, are also direct applications of these laws.

Comment

Although the shortest proof of Nicomachus' theorem is undoubtedly in the proof of the equation

$$y_{harm} * x_{arith} = z_{geom}^2.$$

If there would exist a form, for which can be simply used the formulas for both medieties, then there is besides the preceding proof also an ancient form, which uses only the basic-rules of the laws of calculation for proportions and which does not use the short way by means of fusion: This we want to demonstrate and claim for fun:

$$y_{harm} : z_{geom} : x_{arith} \text{ is a geometric chain of proportions.}$$

For the sake of shorter formula designs we use the symbols $x = x_{arith}$, $y = y_{harm}$, $z = z_{geom}$ and note first for the size arrangement:

Because $a < x < b$ is presupposed, it is then also true that $(a : a) < (a : x) < (a : b)$, from which, according to the definition of the proportions as ratios of $(x - a)$ to $(b - x)$, also follows their corresponding size ratio to each other. Now, we carry out the proof of the equation, namely that

$$y : z \cong z : x$$

holds (which is, after all, the criterion of a geometric-chain of proportions) and this proof we structure into several steps:

1. Assertion: proportion similarity applies $2x : (a + b) \cong a : a$.

Because $a : a \cong x : x$, it follows from $(x - a) : (b - x) \cong a : a$ also $(x - a) : (b - x) \cong x : x$. Now we add these similar proportions with the sum-rule and get $(2x - a) : b \cong x : x$. If we write for $x : x$ again $a : a$, so follows the 1st assertion again after the sum-rule from $(2x - a) : b \cong a : a$.

2. Assertion: proportion similarity applies $y : a \cong 2b : (a + b)$.

For from $(y - a) : (b - y) \cong a : b$ follows with the sum-rule $y : (2b - y) \cong a : b$ and from this with the cross-rule the proportion $y : a \cong (2b - y) : b$ and from this again with the sum-rule that $y : a \cong 2b : (a + b)$ holds.

3. Assertion: proportion similarity applies $y : a \cong b : x$.

For from the assertions 1 and 2 follow with the cross-rule the two similarities $2x : a \cong (a + b) : a$ and also $2b : y \cong (a + b) : a$. Now the factor 2 can be 'shortened', because with the cross-rule applies in general

$$2u : s \cong 2v : t \Leftrightarrow 2u : 2v \cong s : t \Leftrightarrow (2 : 2) \odot (u : v) \cong u : v \cong s : t,$$

because the number proportion $2 : 2$ can only be interpreted as '1'. But another reason is given by the 'difference-rule': u is a piece of the whole $2u$ as well as v of the whole $2v$. The pieces now behave like the wholes (therefore the proportions for the remainders are equal). From $x : a \cong b : y$ follows now according to the cross-rule the 3rd assertion.

4. Assertion: proportion similarity applies $y : z \cong z : x$.

We now get this through the insertion-rules together with the rules for the fusion of proportions:

$$y : z \cong ya : za \cong (y : a) \odot (a : z) \cong (b : x) \odot (a : z) \cong (b : x) \odot (z : b) \cong (z : x)$$

and from this follows the fourth assertion.

Some Comments

1. Musical proportion **(the chain of proportions of the Pythagorean canon)** is thus realized first and foremost in the concrete octave canon $6 - 8 - 9 - 12$:

$$6 : 8 \cong 9 : 12 \quad \text{respectively} \quad 6 : 9 \cong 8 : 12.$$

These proportions regulate the important musical case of the octave $1 : 2 \cong 6 : 12$ and lead to the fifth (9) and the fourth (8). In this the meaning of this proportion equation of Nicomachus becomes visible, of which he says,

it is the most perfect, threefold, all-encompassing equation, which is of the greatest use for every kind of musical or scientific investigation. (cf. [6], p. X).

And Boethius (c. 480–525 AD), the most prominent late-antique scholar of late antiquity, also calls it the *'harmonia perfecta maxima' (cf. [6], p. X)*.

The Neoplatonist Iamblichos (c. 245–325 AD) mentions that this equation was known not only to Pythagoras, but already earlier to the Babylonians (ibid.).

2. We will, however, extend this 'harmonia perfecta maxima', while preserving its symmetries, so that the complete building of the usual third-quint diatonic is obtained from it. In the process, far more than these symmetries come to light.

3. The common use of the terms **'authentic'** and **'plagal'** (sometimes also 'plagalic') in Gregorian chant in connection with the characterization of ecclesiastical-tonal scales does not have anything directly to do with our common major-minor system. However, In only one (perhaps the most obvious) case do we encounter the major-minor opposition as an interval to be decomposed as authentic or plagal. That is the case of the pure fifth:

With the choice of fifth proportion $Q \cong 2 : 3 \cong 20 : 30 \cong a : b$ is $y_{harm} = 24$ and $x_{arith} = 25$ and then the authentic-plagal adjuncts are the two chains

$$20 : 25 \oplus 20 : 24 \cong 20 : 25 \oplus 25 : 30 \cong 20 : 25 : 30 \cong 4 : 5 : 6 \cong 4 : 5 \oplus 5 : 6,$$
$$20 : 24 \oplus 20 : 25 \cong 20 : 24 \oplus 24 : 30 \cong 20 : 24 : 30 \cong 10 : 12 : 15 \cong 10 : 12 \oplus 12 : 15.$$

This corresponds in exemplary fashion to the chord structure of the fifth as a layering of major third (4 : 5) and minor third (5 : 6) in the two possible series – what is known as the major and minor chord of the pure diatonic, that is:

$$\text{major third } (4 : 5) \oplus \text{minor third } (5 : 6) \ (\equiv \text{major} - \text{triad}),$$
$$\text{minor third } (5 : 6) \oplus \text{major third } (4 : 5) \ (\equiv \text{minor} - \text{triad}).$$

4. The juxtaposition of major and minor forms (the pure diatonic) can be seen in yet another context: Namely, we will discover:

▶   that the reciprocal proportion chain of an arithmetic proportion chain is a harmonic one and vice versa.

And as we have just seen, the 'arithmetic' proportion chain 4 : 5 : 6 in pure temperament musically describes the major triad (for example, the C major triad $c - e - g$). A reciprocal is the 'harmonic' proportion chain

$$(1/6) : (1/5) : (1/4) \cong 10 : 12 : 15,$$

and this describes the diatonic pure minor chord (for example $d - f - a$).

The early historical division of numbers into arithmetic and harmonic ones, as we noted in Sect. 1.1 on the occasion of the description of the 'Genesis', is thus obviously consistent with the fact that

- *each three successive members of the natural numbers – called* **perissos numbers** *in ancient Greece – form an* **'arithmetic'** *chain of proportions*

and that then as well

- *three successive magnitudes (members) of reciprocal natural numbers (called* **artios numbers***) form a* **'harmonic'** *chain, which we will show later.*

Here, we may also assume one of the origins of the terms **'arithmetic sequences'** and the like – after all, the natural numbers are the universal model case of such arithmetic regularities. And we could also see the naming **'harmonic sequence'** for the sequence $\left(\frac{1}{n}\right)_{n \in \mathbb{N}}$ anchored in these contexts.

Of course, in addition to these three main forms of medieties, there were many more mean value proportions in antiquity. We will present some of them (among them also those relevant for diatonics) in the next Sect. 3.3.

3.3 The Mean Value Chains of the Ancient Averages

When studying the ancient mean values – "medieties" one comes across hints again and again that besides the Babylonian medieties (geometric, arithmetic and harmonic) there had been a number of other averages. Thus, one finds some hints that besides the geometric mean (whose numerical magnitude was treated as *'unnameable'* or even as *'unpresentable'*, because of its almost constant irrational value) there must have been another nine medieties. In the following Definition 3.2 we now take up both the already known Babylonian averages and add to them their so-called 'contra-medieties'. Here are added further averages, to which the designation 'homothetic' was given.

In all descriptions, we use the uniform **'piece to the rest'** formulation, which we also provide in the two **'piece to the whole'** forms for the convenience of the reader,

$$(x - a) : (b - x) \cong \ldots \text{(piece-to-the-rest-form)}$$
$$(x - a) : (b - a) \cong \ldots \text{ and } (b - x) : (b - a) \cong \ldots \text{(piece-to-the-whole-forms)}.$$

The reason for this is that the proportions mentioned in the literature occur in different forms (even within the same reading) which makes a comparison with each other as well as an overview of them difficult. The geometric mean has, besides its main form (square-rectangle-form), many forms, which are all equivalent possibilities for definition.

Definition 3.2 (The Ancient Medieties and their Proportion Chains)
Let $a : x : b$ be a 3-membered (number) proportion chain, of which we assume the
ascending order $a < x < b$. The **proportion chain** is then called

1. **Geometrically** $\Leftrightarrow a : x \cong x : b$ **(Square-Rectangle Formula)**,
    * $(x - a) : (b - x) \cong a : x$ and $(x - a) : (b - x) \cong x : b$,
    * $(x - a) : (b - a) \cong a : (a + x)$ and $(x - a) : (b - a) \cong x : (b + x)$.
    For the geometric chain of proportions, two contra-geometric medieties can be
    defined at once: The proportion chain is called
2. **Contra-Geometric (I)** $\Leftrightarrow (x - a) : (b - x) \cong x : a$,
    * $(x - a) : (b - a) \cong x : (a + x)$ and $(b - x) : (b - a) \cong a : (a + x)$,
3. **Contra- Geometric (II)** $\Leftrightarrow (x - a) : (b - x) \cong b : x$,
    * $(x - a) : (b - a) \cong b : (b + x)$ and $(b - x) : (b - a) \cong x : (b + x)$.
Four non-geometric mean forms follow, which in our context are also called
musical medieties. The proportion chain is called
4. **Arithmetic** $\Leftrightarrow (x - a) : (b - x) \cong a : a \cong 1 : 1$,
    * $(x - a) : (b - a) \cong a : (a + a) \cong 1 : 2$ and $(b - x) : (b - a) \cong 1 : 2$,
5. **Contra-Arithmetic** $\Leftrightarrow (x - a) : (b - x) \cong a^2 : b^2$,
    * $(x - a) : (b - a) \cong a^2 : (a^2 + b^2)$ and $(b - x) : (b - a) \cong b^2 : (a^2 + b^2)$,
6. **Harmonic** $\Leftrightarrow (x - a) : (b - x) \cong a : b$,
    * $(x - a) : (b - a) \cong a : (a + b)$ and $(b - x) : (b - a) \cong b : (a + b)$,
7. **Contra-Harmonic** $\Leftrightarrow (x - a) : (b - x) \cong b : a$,
    * $(x - a) : (b - a) \cong b : (a + b)$ and $(b - x) : (b - a) \cong a : (a + b)$.
The next four (**homothetically** named) medietary forms divide the distance $(b - a)$
in such a way that it is in proportion to the vertices a and b. Thus, a chain of
proportions is called
8. **Subhomothetic** $\Leftrightarrow (x - a) : (b - x) \cong a : (b - a)$,
    * $(x - a) : (b - a) \cong a : b$ and $(b - x) : (b - a) \cong (b - a) : b$,
9. **Contra-Subhomothetic** $\Leftrightarrow (x - a) : (b - x) \cong (b - a) : a$,
    * $(x - a) : (b - a) \cong (b - a) : b$ and $(b - x) : (b - a) \cong a : b$,
10. **Superhomothetic** $\Leftrightarrow (x - a) : (b - x) \cong b : (b - a)$,
    * $(x - a) : (b - a) \cong b : (2b - a)$ and $(b - x) : (b - a) \cong (b - a) : (2b - a)$,
11. **Contra-Superhomothetic** $\Leftrightarrow (x - a) : (b - x) \cong (b - a) : b$,
    * $(x - a) : (b - a) \cong (b - a) : (2b - a)$ and $(b - x) : (b - a) \cong b : (2b - a)$.
Designations: For the **mean values x (medieties),** which satisfy the corresponding
proportions, we use – depending on the context – two designation models: On the
one hand, this is an indexed listing obeying the preceding order.

x_I for the geometric medium, ..., x_{XI} for the contra - superhomothetic medium,

(continued)

Definition 3.2 (continued)

on the other hand, these are symbols provided by an explanatory subscript, which, especially in the case of the geometrical and the musical medieties, bear a more concise designation pattern: First of all, the geometrical medieval (of two fixed dates a and b) is usually represented by the symbol

$$z_{geom}(a, b) \text{ or rather only briefly } z_{geom}$$

and the musical medieties are predominantly given the symbolism

$$x_{arith} \text{ (for arithmetic) and } x_{co\text{-}arith} \text{ (for contra-arithmetic)},$$
$$y_{harm} \text{ (for harmonic) and } y_{co\text{-}harm} \text{ (for contra-harmonic)}.$$

All homothetic medieties receive the variable u with corresponding subscript.

On these terms, let us first note the following:

1. All medians in our list –with the provisional exception of the geometric mean (and its two contra-forms) – describe at first sight the relation of the two partial distances $(x - a)$ and $(b - x)$ of the total distance $(b - a)$ to each other, where the proportions depend exclusively on the given data a and b and not on the division point (x) itself. And we also see that the division parameters can all be expressed as functions of the frequency measure $q = b/a$, as already indicated in Theorem 3.1. Moreover, we see: The proportions of the divisions with respect to each other are similar to the proportions of the distance ratios that can also be interpreted by the data a and b. For example, in the case of the 'homothetic' mediety, the division of the total distance $(b - a)$ into the two partial distances $(x - a)$ and $(b - x)$ is carried out similarly ('proportionally') to the division of the total distance (b) into its partial distances (a) and $(b - a)$ (which also explains the naming).
2. However, beyond what was said in (1), we note that the medietary proportions also partly take the concise form of

$$(x - a) : (b - a) \cong q^n : 1 \cong b^n : a^n,$$

where $q = b/a$ is the frequency measure of the interval $[a, b]$, and n is an exponent which is an integer (at least in the non-geometric means). For the geometric mean itself, $n = -(1/2)$: Indeed, the following theorem can be shown for the piece-at-rest form:

Theorem The geometric mediety can also be characterized in a piece-to-rest proportion beyond the forms given in Definition 3.2. Furthermore, it follows that the division parameter also depends only on the quotient $q = b/a$:

$$a : x \cong x : b \Leftrightarrow (x - a) : (b - a) \cong \sqrt{a} : \sqrt{b} = q^{-1/2}.$$

3. With the exception of the arithmetic mean, the inversion of the division ratio which the two parts $(x - a)$ and $(b - x)$ of the total distance $(b - a)$ have to each other leads to the mean values known as 'contra-medians'. However, the arithmetic case can be explained in the context of a different system. We refer to the following remark (4) as well as to the later Sect. 4.2 about the contra-medietary sequences.

4. A mathematically supported view about these terms

'Mediety – Contra-Mediety'

and their interpretations. We obtain the **frequency measure symmetry** from Theorem 3.1 and the functional equation there, and then we obtain the observation expressed in a theorem:

Theorem Let x_1 be a medietary with the piece-to-rest partition parameter $y = \varphi(q)$. Then let x_2 be another medietary with the piece-to-rest-partition parameter

$$\psi(q) = 1/q\varphi(q).$$

So that means that the piece-to-rest proportions

$$(x_1 - a) : (b - x_1) \cong \varphi(q) \quad \text{and} \quad (x_2 - a) : (b - x_2) \cong \psi(q) = 1/q\varphi(q)$$

exist. Then applies

$$x_2^* = x_1 \text{ respectively } x_1 x_2 = ab.$$

Thus, the two medieties x_1, x_2 lie mirrored, both to the edges a and b and to the geometric mean z_{geom}. Thus, the coordinate point (x_1, x_2) lies on the hyperbola of the archytas $y = ab/x$.

So specifically, the medietes to the division parameters $\varphi(q) = q^n$ and $\psi(q) = q^{-(n+1)}$ are in mirrored position.

Unfortunately, the ancient ways of designation do not follow this picture, which in our opinion would possess a more consistent 'logic'.

5. Based on the foregoing, we see that all medieties could be constructed exclusively by means of the ray theorems of elementary geometry.

The geometric mean is also very easy to construct. Geometrically, if we use the square-rectangle formula of proportion $a : x \cong x : b$ as the equation

$$x^2 = ab$$

and then use the 'cathetus theorem' (as well as the 'height theorem') from the Pythagorean theorem group. It should be noted that these two theorems are also consequences of the ray theorems.

The following example package first shows us a whole series of concrete numerical ratios for all these averages:

Example 3.2 Ancient Medieties

In the table decimals with punctuation are rounded data.

Mediety	Proportion integer	Proportion with $a = 6$	Octave $6 : x : 12$	Fifth $6 : x : 9$
x_I – geometrical	$1 : 2 : 4$	$6 : 12 : 24$	$6 : 8, 48. : 12$	$6 : 7, 35. : 9$
x_{II} – contra-geometric I	$2 : 4 : 5$	$6 : 12 : 15$	$6 : 9, 70. : 12$	$6 : 7, 68. : 9$
x_{III} – contra-geometric II	$1 : 4 : 6$	$6 : 24 : 36$	$6 : 9, 36. : 12$	$6 : 7, 62. : 9$
x_{IV} – harmonic	$3 : 4 : 6$	$6 : 8 : 12$	$6 : 8 : 12$	$6 : 7, 2 : 9$
x_V – contra-harmonic	$3 : 5 : 12$	$6 : 10 : 12$	$6 : 10 : 12$	$6 : 7, 8 : 9$
x_{VI} – arithmetical	$2 : 3 : 4$	$6 : 9 : 12$	$6 : 9 : 12$	$6 : 7, 5 : 9$
x_{VII} – contra-arithmetic	$5 : 6 : 10$	$6 : 7, 2 : 12$	$6 : 7, 2 : 12$	$6 : 6, 92. : 9$
x_{VIII} – subhomothetic	$6 : 8 : 9$	$6 : 8 : 9$	$6 : 9 : 12$	$6 : 8 : 9$
x_{IX} – contra-subhomothetic	$3 : 7 : 9$	$6 : 14 : 18$	$6 : 9 : 12$	$6 : 7 : 9$
x_X – superhomothetic	$5 : 11 : 15$	$6 : 13, 2 : 18$	$6 : 10 : 12$	$6 : 8, 25 : 9$
x_{XI} – contra-superhomothetic	$5 : 9 : 15$	$6 : 9 : 10, 5$	$6 : 8 : 12$	$6 : 6, 75 : 9$

◄

In the music-theory of pure diatonics, these four medieties are the most common,

- the harmonic and the arithmetic means,
- the contra-harmonic and the contra-arithmetic means

of our listing. We can see from the above Example 3.2, that in the case of the octave $6 : 12$ as total proportion, all the homothetically named mean values would not bring about any new divisions. The geometric mean is here only indirectly in the boat. On the other hand, in other considerations, the other medieties (and a thousand others) could also appear. The play with the higher proportions (presented in Chap. 4), as well as in the next Sect. 3.4 shows us a never-ending wealth of possibilities for generating ever newer medieties from the existing ones.

The next example illustrates some basic connections between mean-values (medieties), proportions and music:

Example 3.3 Musical Medieties and Their Chains of Proportions

1. The **Geometric Proportion Chain** $3 : 6 : 12$ (chord example $d - d' - d''$).
2. The **Arithmetic Proportion Chain** $6 : 9 : 12$ (chord example $d - a - d'$).
3. The **Harmonic Proportion Chain** $6 : 8 : 12$ (chord example $d - g - d'$).

   Pythagorean music theory developed from the internal structures of the musical chain of proportions B_{mus}

$$B_{mus} = a : y_{harm} : x_{arith} : b \cong 6 : 8 : 9 : 12,$$

and the whole chain corresponds. For example, to the tone sequence or the chord

$$d - g - a - d'$$

in Pythagorean pure fifth temperament. In the later diatonic, which includes the pure third $4 : 5$, the two contra-medieties of arithmetic and harmonic mean also play a decisive role. We have

4. the **Contra-Harmonic Proportion Chain** $6 : 10 : 12 \cong 3 : 5 : 6$,

   which corresponds to a sequence of notes $d - h - d'$ – or also the composition of a minor pure third $5 : 6$ to a major pure sixth $3 : 5$. We also have

5. the **Contra-Arithmetic Proportion Chain** $6 : 7, 2 : 12 \cong 5 : 6 : 10$,

   which corresponds to a minor chord $d - f - d'$ or $a - c - a'$ within a scale built up by pure fifths $(2 : 3)$ and pure thirds. $(4 : 5)$

Abstract: The medieties thus obtained can be arranged in series thus:

$$6 - 7, 2 - 8 - 9 - 10 - 12,$$

which is the chain of proportions, which we also formulate as a similar chain in integer form,

$$a : x_{co\text{-}arith} : x_{harm} : x_{arith} : x_{co\text{-}harm} : b \cong 30 : 36 : 40 : 45 : 50 : 60,$$

corresponds. We can verify this on the 'white keys' of a keyboard that is in fact tuned purely diatonically by the tone sequence

$$d - f - g - a - h - d'.$$

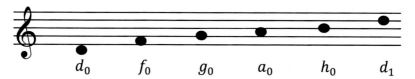

With the exception of the semitonia, which are interval differences, this chain already contains all the steps and structural intervals of the 'pure diatonic' scale:

- $30 : 36 \cong 5 : 6 \equiv$ minor third
- $36 : 40 \cong 9 : 10 \equiv$ small whole tone
- $40 : 45 \cong 8 : 9 \equiv$ great whole tone
- $40 : 50 \cong 36 : 45 \cong 4 : 5 \equiv$ major third
- $30 : 40 \cong 3 : 4 \equiv$ perfect fourth
- $30 : 45 \cong 2 : 3 \equiv$ perfect fifth
- $30 : 50 \cong 3 : 5 \equiv$ major sixth
- $30 : 60 \cong 1 : 2 \equiv$ Octave.

If we form the pure fourth $30 : 40 \cong 36 : 48$ as well as the pure major third in the form $36 : 45$, the proportion chain $36 : 45 : 48$ directly yields the difference interval 'fourth minus major third'

- $45 : 48 \cong 15 : 16 \equiv$ pure diatonic semitone.   ◀

In these listed chains of musical proportions, two more regularities (which at first seem rather coincidental) catch our attention:

- First, the chains in (2) and in (3) are reciprocal.

However, after reading the Harmonia perfecta maxima of the Babylonian medieties (Theorem 3.2), we are familiar with this: Arithmetic mean and harmonic mean are mirrored and define the symmetrical octave canon. What is surprising, however, is that also

- second, the chains in (4) and in (5) are reciprocal,

at least this is true here in the case of the octave. How this is in general, we will see yet. In the meantime, we see from this that already the 3-step-chain

$$6 : x_{\text{arith}} : y_{\text{co-harm}} : 12 = 6 : 9 : 10 : 12$$

would be sufficient to build the quint-terz structure – called **pure diatonic.**

Another example block is dedicated to the geometric mean:

Example 3.4 Geometric Averaging

(A) For $a^2 = 16$ and $b^2 = 25$ ist $4*5 = 20$ the geometric mean of these two squares and $16 : 20 : 25$ is a geometric proportion chain.

Geometrically: The area of the square with side length 20 is equal to the area of the rectangle with sides 25 and 16.

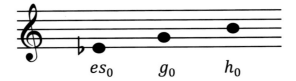

$$es_0 \qquad g_0 \qquad h_0$$

Musically: The proportion chain $16 : 20 : 25$ is the representation of the superposition ('composition') of two perfect major thirds $4 : 5$, and this would correspond, for example, to the tone sequences $es - g - h$ or $c - e - gis$ – provided that the pure tuning were continued into the chromatic.

(B) With the composition of two intervals of the same size – or better: Two similar intervals – a geometrical proportion chain always arises: If $A = a : b$ and $C = c : d$, then after the construction of the composition the following holds true

$$[a, b] \oplus [c, d] \cong [ac, bc] \oplus [cb, db] \cong ac : bc : db,$$

and then is $ac : bc = a : b$ and $bc : db = c : d$. Is now

$$a : b \cong c : d, \text{so is } ac : bc \cong bc : db,$$

and the chain is geometric by definition, which of course is shown by the square-rectangle equation: From $bc = ad$ follows $(bc) * (bc) = (bc) * (ad) = (ac) * (db)$.

(C) A wide field of geometric proportion chains is consequently found in the **Equal-Step Temperings** (precisely because the equal step is achieved precisely according to the geometric proportions). ◀

Next we want to give (in the context of a usual equation treatment) the mean formulas which can be obtained very easily from the proportions, as well as some properties which follow from them:

Theorem 3.3 (The Mean Value Formulas and the Mean Value Ordering)

(A) **Mean value formulas**

All medians can be conveniently determined as unique solutions of the mean equations or thanks to the mean function and the details are tabulated these:

Proportion name	Division parameters – piece-to-rest prop $(x - a) : (b - x) \cong$	Mean value formula
x_I – geometrical	$\sqrt{a} : \sqrt{b}$	$x_I^2 = ab$
x_{II} – contra-geometric-I	$x : a$	$x_{II} = \frac{b-a}{2} + \sqrt{a^2 + \left(\frac{b-a}{2}\right)^2}$
x_{III} – contra-geometric-II	$b : x$	$x_{III} = \frac{a-b}{2} + \sqrt{b^2 + \left(\frac{b-a}{2}\right)^2}$
x_{IV} – arithmetical	$1 : 1$	$x_{IV} = \frac{1}{2}(a + b)$
x_V – contra-arithmetic	$a^2 : b^2$	$x_V = \frac{ab}{a^2+b^2}(a + b)$
x_{VI} – harmonic	$a : b$	$x_{VI} = \frac{2ab}{a+b}$
x_{VII} – contra-harmonic	$b : a$	$x_{VII} = \frac{a^2+b^2}{a+b}$
x_{VIII} – subhomothetic	$a : (b - a)$	$x_{VIII} = \frac{a(2b - a)}{b}$
x_{IX} – contra-subhomothetic	$(b - a) : a$	$x_{IX} = \frac{b^2+a(a-b)}{b}$
x_X – superhomothetic	$b : (b - a)$	$x_X = \frac{b^2+a(b-a)}{2b-a}$
x_{XI} – contra-superhomothetic	$(b - a) : b$	$x_{XI} = \frac{b^2}{2b-a}$

(B) **The arrangement of the musical mean values to the Diatonic proportion chain**

The musical medieties fulfill the inequalities

$$a < x_{\text{co-arith}} < y_{\text{harm}} < (z_{\text{geom}}) < x_{\text{arith}} < y_{\text{co-harm}} < b,$$

and from this arises the 5-step **Diatonic musical proportion-chain**

$$M_{\text{diat}} = a : x_{\text{co-arith}} : y_{\text{harm}} : x_{\text{arith}} : y_{\text{co-harm}} : b.$$

It is the main subject of the Harmonia perfecta maxima diatonica.

Proof: For the mean value formulas one solves the corresponding equation according to the variable x. Undoubtedly, there would also be a form of proof based on a pure calculus with proportions (also in the case of the root expressions). For example, in the contra-harmonic case, one would prove the mean by the proportion relation

$$x_{VII} : 1 \cong (a^2 + b^2) : (a + b)$$

and a derivation would transfer the defining proportion of the contra-harmonic into this form by means of the calculation-rules of the theory of proportions.

What remains is the insight into how, in the case of the geometric mean, one can get at the shape of the division parameter

$$f(x_I) = f(z_{geom}) = \sqrt{a/b}$$

comes. If we look carefully at the Example 3.1, this is actually already clear. After all, we use there the functional equation of the mean value function $g(y)$. But it also works like this: Because it is clear that $x_I^2 = ab$ is (this follows immediately from the most important definitional description of the geometric mediety $a : x_I \cong x_I : b$), we simply put this value into one of the two piece-of-residue-forms, we get

$$f(x_I) = \frac{a}{x_I} = \frac{a}{\sqrt{ab}} = \frac{\sqrt{a}}{\sqrt{b}} = \sqrt{a/b} = \frac{x_I}{b},$$

and this proof is already given. For the contra-geometric medieties we save a representation of the division parameters $f(x_{II})$ and $f(x_{III})$ as a function of the data a and b. For this one must insert only the respective mediety-formula into the division parameter.

To (B): Even if the chain of inequalities could be obtained from the mean value formulas via extensive computations, it follows directly from the appropriate interpretation of the definition of proportions: For there the size relation of the sections $(x - a)$ and $(b - x)$ to each other is described, and then we can apply the mean value monotonicity from Theorem 3.1.

Note: The chain of inequalities of the Babylonian medieties themselves (known as the 'inequality between harmonic and arithmetic mean') can otherwise be recognized in many ways, one of which may serve as a demonstration:

$$0 < (b - a)^2 = b^2 + a^2 - 2ab$$
$$\Leftrightarrow 2ab < b^2 + a^2 \Leftrightarrow 4ab < b^2 + a^2 + 2ab$$
$$\Leftrightarrow z_{geom}^2 = ab < \frac{1}{4}(a^2 + 2ab + b^2) = (\frac{1}{2}(a + b))^2 = x_{arith}^2.$$

Thus the inequality $z_{geom} < x_{arith}$ is shown. But because the product of y_{harm} with x_{arith} is equal to $(z_{geom})^2$, then the other factor y_{harm} must be smaller than the geometric mean. z_{geom}

We now give some useful observations from these formulas:

Corollaries: Equations, Inequalities, Mediate Order

1. If the magnitudes a and b *are* rational, then with the exception of the geometric means, which can then also be irrational (and 'usually' are), all other mediations are rational. Thus, they are commensurable with each other.
2. Simultaneously, the **arithmetic mean** is also the arithmetic mean of the harmonic and contra-harmonic mean:

$$x_{arith} = \frac{1}{2}\left(y_{harm} + y_{co\text{-}harm}\right),$$

because we have the same differences

$$x_{arith} - y_{harm} = \frac{1}{2}\frac{(a-b)^2}{a+b} = y_{co\text{-}harm} - x_{arith}.$$

3. Simultaneously, the **geometric mean** of two data a and b is also the geometric mean for their harmonic and arithmetic mean as well as for the contra-arithmetic and contra-harmonic mean, namely the formulas

$$y_{harm} * x_{arith} = x_{co\text{-}arith} * y_{co\text{-}harm} = ab = \left(z_{geom}\right)^2.$$

Whereby the list of all relevant properties around the geometric mean contains quite other and more profound aspects (see the following Sect. 3.4).

4. The **harmonic mean** always satisfies the boundary condition with respect to its position

$$a < y_{harm} < 2a,$$

and if b *is* monotonically increasing from a to $+\infty$, then $y_{harm}(a, b)$ is monotonically increasing from a to $2a$. We read this quite easily from the form

$$y_{harm}(a, b) = \frac{2ab}{a+b} = 2a\left(\frac{b}{a+b}\right) = 2a\left(\frac{1}{1+a/b}\right),$$

because the fraction is obviously always smaller than 1 and it tends monotonically increasing towards 1, if b grows and becomes unrestrictedly large compared to a, so that therefore a/b tends towards 0. But already the definition of these proportions also shows us these conditions of position: The proportion of the harmonic mean

$$(y_{\text{harm}} - a) : (b - y_{\text{harm}}) \cong a : b$$

is due to the cross-rule equivalent to the proportion

$$(y_{\text{harm}} - a) : a \cong (b - y_{\text{harm}}) : b.$$

If we imagine the magnitudes as distances, it is clear that the right proportion is numerically smaller than 1, because $(b - y_{\text{harm}})$ is a partial distance of the total distance b. Accordingly, the distance $(y_{\text{harm}} - a)$ is also smaller than the distance a, which is why y_{harm} must satisfy the position condition $a < y_{\text{harm}} < 2a$. So much for the plausibility of this.

5. Simoultaneously, the **harmonic mean** y_{harm} is also the harmonic mean of the contra-arithmetic and arithmetic mean, because the quotient formulas are valid

$$\frac{y_{\text{harm}} - x_{\text{co-arith}}}{x_{\text{arith}} - y_{\text{harm}}} = \frac{2ab}{a^2 + b^2} = \frac{x_{\text{co-arith}}}{x_{\text{arith}}},$$

and that is itself again the defining fractional-arithmetic written piece-at-rest form for the magnitude y_{harm}, being the harmonic mean of the magnitudes $x_{\text{co - arith}}$ and. x_{arith}

6. For the **harmonic mean of** two data a and b we recognize the alternative definition possibility as '**reciprocal of the arithmetic mean of the reciprocals**':

$$y_{\text{harm}} = \left[\frac{1}{2} \left(\frac{1}{a} + \frac{1}{b} \right) \right]^{-1}.$$

There is also another extremely useful form: For example, if you want to avoid a formula, describe the **geometric mean** of two dates a and b as the side length of a square equal in size to the rectangle of sides a and b and with the **harmonic mean**, this works almost as elegantly:

▶   *We replace the rectangle (of sides a and b) by **a rectangle of equal size**, where one side is supposed to be the **arithmetic mean** of a and b – the other is then in fact automatically the (sought) **harmonic mean**.*

This can be seen in the formula

$$y_{\text{harm}} = ab/x_{\text{arith}}$$

for the harmonic mean. So one simply forms the product (of a and b) and divides it by the arithmetic mean (child's play and quite equal to the TR in the case of decent numerical specifications by mental arithmetic) and a comparatively laborious calculation from the proportion-rule is hopelessly inferior here. By the way, the same formula proves the mirror property just as drastically: Both data (x_{arith}, y_{harm}) lie on the hyperbola of Archytas.

How memorable this 'averaging procedure' is with regard to the task of converting a rectangle into a square with the same area and how easy it is to obtain the harmonic mean, may be demonstrated by a simple example: Thus, the rectangle consisting of sides 6 and 12 with area 72 transforms in the first step into a much more 'square-like' rectangle: The sides are the arithmetic mean 9 and $8 = 72/9$ – the harmonic mean. Now what results if we again make this new rectangle even more square using the same procedure? Well, the new arithmetically averaged side is $17/2 = 8.5$, and consequently we have the harmonic partner with $72/8, 5 = 144/17$ right in front of our eyes. It comes out 8.470588. . .right in the pocket calculator.

▶   **Important**

But after the occupation with proportions has made us ardent admirers of fraction arithmetic and its commensurability and after we have mastered many an art of similarity arithmetic by mental arithmetic, we despise mere numerics for this time as well and continue to remain faithful to proportions: In the completely new rectangle, the side proportion is namely this:

$$\frac{144}{17} : \frac{17}{2} \cong \frac{288}{34} : \frac{289}{34} \cong 288 : 289.$$

Apparently you have to be a very obsessive accuracy fanatic to not see a square in such a proportioned rectangle.

This example will also serve us in later Sect. 4.3 as a demonstration of the **superfast approximation** of the geometric mean by the **nested mediate sequence** of arithmetic and harmonic means.

3.4   **Games with Babylonian Proportion-Chains**

To anticipate: Anyone who makes the attempt to study the ancient musical world structure in the likewise already historical literatures of the music theorists will find himself visibly in a situation like someone who is supposed to find his way through a Chinese metropolis without a navigation system: many lanes (many branches) foreign languages (foreign signs) and in general! Thus, for example, in the volumes by Ambros and in those by

Freiherr von Thimus, we learn in hundreds of ramifications how the ancient tonal system may have developed and built itself up under the laws of the proportions of the 'ancients'. A taste?

> ...the play of the interpolating medieties can be contrasted, as a contrast to it, for the discovery of new sound levels, with the play of the third proportions of new front or rear members to be added to the proportions to be formed from the functions of new values α and ω of the two end tones of a musical interval for the comparison of their oscillation masses or wavelengths. In addition to α and ω, the third geometric proportional is the front element $\frac{\alpha^2}{\omega}$ and, in relation, the rear element $\frac{\omega^2}{\alpha}$. The values $2\alpha - \omega$...are found as arithmetic front element and rear element (...). (From: von Thimus, Vol I Preface XV)

In this section we will introduce the essentials around these concepts: The subject matter is relationships between geometric, arithmetic and harmonic chains of proportions. And we divide our section into two parts: First, we focus on a play of smaller chains with so-called **third proportions**. It is followed by a fundamentally valid symmetry of the **arithmetic** and the **harmonic proportion-chains** involving their reciprocals.

A separate section (the following Sect. 3.5) is then deservedly devoted to **geometric mediety**. Furthermore, our considerations lead to an **infinite** iteration process of continuing averages (the subject of the last mathematical Chap. 4).

Before that, we remind again of the notions of Babylonian chains: An initially merely 2-step proportion chain is called geometric/arithmetic/harmonic, if the middle member is the geometric/arithmetic/harmonic mediate of the outer members (just as this is stated in the central Definition 3.2). These terms would be transferable in an analogous way to the other medieties, if desired.

(A) The 'third proportionals'

In an averaging of a given interval $[a, b]$ (or proportion $a : b$), we have sought a value x – the **mean proportional** – such that the proportion chain $a : x : b$ is of a very specific given type – e.g. geometric, harmonic, etc., or such that the variable x is one of the classical mean values. However, one could just as well demand that either.

- $x < a < b$ is to be determined in such a way that now the **magnitude** a is the mean value of the magnitudes x and b of the required type. Respectively it is required, that the proportion chain $x : a : b$ is a desired mean value proportion chain,

or that

- $a < b < x$ is to be determined in such a way that now the **magnitude** b is the mean value of the magnitudes a and x of the required type – respectively, that the proportional -chain $a : b : x$ is a desired mean value proportion chain.

In the first case x is called a **'front third proportional'**, in the second case a **'back third proportional'**. We can certainly easily calculate these 3rd proportionals from the data (a, b) by using the mean value formulas of Theorem 3.3.

It turns out that in this case a task tailored to the point and its solution provide the necessary clarity in many computational, but also theoretical considerations. That is the simple

Problem For a given proportion $a : x$, find a front harmonic 3rd proportional y – respectively: Compute a magnitude y such that the 2-step proportion chain $y : a : x$ is harmonic.

Here, we have used the external magnitude x as a variable parameter, and the magnitude y to be searched for is then a function of x, treating the magnitude a as constant. Surely, the solution is quite simple:

Solution The proportion chain $y : a : x$ is harmonic if and only if a is the harmonic mean of y and x, and then the mean formula of Theorem 3.3 applies to a, which we simply rearrange to y and then still represent the result proportionally. In formulas this process reads like this:

$$a = \frac{2yx}{y+x} \Leftrightarrow y = \frac{ax}{2x-a} \Leftrightarrow y : a \cong x : (2x-a) \cong 1 : \left(2 - \frac{a}{x}\right).$$

This fractional-rational function

$$y = y(x) = \frac{ax}{2x-a}$$

is of the geometric type of a hyperbola – the **'harmonic hyperbola'** and allows a convenient progression analysis, which we also find sketched in the appendix.

Conclusion If x strives monotonically increasing from a to ∞, then y strives monotonically decreasing and asymptotically from a to $a/2$ – in accordance with our knowledge that the harmonic mean (here a) can never be larger than twice the forelimb (y).

In the next example block, we'll practice these things a bit first:

Example 3.5 'Third Proportionals'

For the parameters a and b in this example, always assume the case of a large pure sext.
$a : b \cong 6 : 10$

Required type for 3rd proportional	x value	Proportion chain
Front arithmetic 3. proportional	$x = 2$	$2 : 6 : 10$
Rear arithmetic 3. proportional	$x = 14$	$6 : 10 : 14$
Front geometric 3. proportional	$x = 3, 6$	$3, 6 : 6 : 10 \cong 9 : 15 : 25$
Rear geometric 3. proportional	$x = 16 \frac{2}{3}$	$6 : 10 : 16\frac{2}{3} \cong 18 : 30 : 50$
Front harmonic 3. proportional	$x = 4\frac{2}{7}$	$4\frac{2}{7} : 6 : 10 \cong 30 : 42 : 70$
Rear harmonic 3. proportional	$x = 30$	$6 : 10 : 30$
Rear contra-harmonic 3. proportional	$x = 12$	$6 : 10 : 12$
Front contra-arithmetic 3. proportional	$x = 5$	$5 : 6 : 10$

◀

For example, in the fifth example, in the case of an anterior harmonic 3rd proportional, one arrives at the given x-value according to our preceding discussion (i.e., in the instantaneous variable terms) by the following reasoning:

$$a = \frac{2xb}{x + b} \Leftrightarrow x = \frac{ab}{2b - a} \Leftrightarrow x = \frac{60}{20 - 6} = \frac{30}{7} = 4\frac{2}{7} \approx 4,285\ldots$$

The proportion chain $30/7 : 6 : 10$ finally changes into the equivalent proportion chain in the integer form $30 : 42 : 70$ by multiplication with 7.

▶ However, such 3rd proportion numbers do not have to exist at all! So there is no rear harmonic 3rd proportional x to the proportion $1 : 3$, because in the chain $1 : 3 : x$ the number 3 would otherwise be the harmonic mean of 1 and x, but a harmonic mean is always smaller than the double of the front magnitude (1) – as already often observed, last in the conclusion 4) according to Theorem 3.3. Also, there is neither a front contra-harmonic nor a rear contra-arithmetic 3rd proportional in the above Example 3.5.

About the possibilities of their existence and the methods to generate such third proportionals in a recursive process is reported in the next Chap. 4, where we want to continue the processes of a mean value generation **ad infinitum,** *'so that mathematics also comes into its own'.*

Now let us anchor these things in a simple and obvious definition:

Definition 3.3 (Babylonian Proportion Chains and Higher Proportions)
Let be given for $n \geq 2$ an n-step proportion chain

$$A = a_0 : a_1 : \ldots : a_n$$

with the n step proportions $a_k : a_{k+1}, (k = 0, \ldots, n-1)$. Then the chain A is called
Geometric/Arithmetic/Harmonic – in short: **Babylonian**
   $\Leftrightarrow$ each of its $(n-1)$ possible 2-level subchains in the sequence of directly adjacent levels

$$a_k : a_{k+1} : a_{k+2}, \ (k = 0, \ldots, n-2)$$

is geometric/arithmetic/harmonic. In connection with these progressive sequences of 2-step proportion chains the magnitudes are called

$$a_k \ front \ \text{3rd} \ Proportional \ of \ a_{k+1} : a_{k+2},$$
$$a_{k+1} \ middle \ \text{3rd} \ Proportional \ of \ a_k : a_{k+2},$$
$$a_{k+2} \ rear \ \text{3rd} \ Proportional \ of \ a_k : a_{k+1},$$

where a characterizing Babylonian mediate property (geometric or arithmetic or harmonic) is added if the corresponding substring has that property.

   This process can be continued and generalized: If a 2-step proportion chain $a : b : c$ is Babylonian, then let x be a magnitude such that the chain is

$$a : b : c : x$$

is also Babylonian. Then x is called a posterior Babylonian **4th Proportional** and completely analogous are all so-called front and rear (Babylonian) **Higher Proportional**.

   Accordingly, one would also define proportion chains with other medietic properties (homothetic, contra-harmonic, etc.).

As a simple example of Babylonian proportion-chains, consider the octave canon $6 : 12$, for which we calculate various front, middle, and back higher proportions (they are printed in bold in the table of the following Example 3.6). Thus, the 1-step proportion $6 : 12$ gives rise to several multi-step Babylonian proportion chains.

Example 3.6 Babylonian Proportion-Chains

Proportion chain	Babylonian type	Higher Proportional
3 : 6 : 12 : 24 : 48 : 96	Geometric	**96** = rear **geom. 6. Prop.**
6 : 12 : 18 : 24	Arithmetic	**24** = rear **arithm. 4. Prop.**
2, 4 : 3 : 4 : 6 : 12	Harmonic	**2, 4** = front **harm. 5. Prop.**

◄

Now one of the most interesting games with Babylonian proportions is in the variety of internal symmetries between chains of different Babylonian types and their interactions with **higher proportionals** and **reciprocal processes.** Let us consider an example:

Example 3.7 Full Diatonic Canon

In the complete diatonic canon

$$6 : 7, 2 : 8 : 9 : 10 : 12$$

let us consider the whole-tone proportion $a : b \cong 8 : 9$. For them $x = 10$ is a back arithmetic 3rd proportional. Likewise it is easy to see that $y = 7, 2$ is a front harmonic 3rd proportional, for magnitude 8 is the harmonic mean of 7.2 and 9, as the calculus

$$8 = 2 * 9 * 7, 2 / (9 + 7, 2),$$

but also the formula for the front harmonic 3. proportional

$$y = \frac{ab}{2b - a} = \frac{72}{18 - 8} = 7, 2$$

show. We observe that in the 3-step proportion chain $P = 7, 2 : 8 : 9 : 10$

the rear partial chain $H = 8 : 9 : 10$ arithmetic,

the front partial chain $V = 7, 2 : 8 : 9$ harmonic

is. Also it is quickly clear, that the chain V is a reciprocal of H – as of course vice versa. And the whole chain P is symmetrical or reciprocal to itself.   ◄

This and similar examples of the interaction of anterior and posterior proportions are encountered at every turn in early music-historical practice, where interval constructions are often carried out in this manner.

Next, we want to transfer the situation of this example to a somewhat more general level (quite simply, because we then arrive at a view that is directed to the core of the content-related connections). For this purpose, we have the following **situation:** Given are the four data

$$x_1 < x_2 < x_3 < x_4,$$

where we assume the ascending order only for convenience. They form the proportion chain

$$P \cong x_1 : x_2 : x_3 : x_4,$$

and we regard the two inner magnitudes x_2 and x_3 as magnitudes of a given proportion or as data of a musical interval $[x_2, x_3]$.

Then we study the interplay of the outer members x_1 and x_4 in connection with certain properties of the entire 3-step proportion chain P and its front and back 2-step subchains V and H, and we then formulate a result in the following proposition:

Theorem 3.4 (Symmetry Game of the Babylonian 3rd Proportional)

Let $P = x_1 : x_2 : x_3 : x_4$ be a 3-step ascending proportion chain with the front and the back 2-step subchain

$$V = x_1 : x_2 : x_3 \text{ and } H = x_2 : x_3 : x_4.$$

1. The following relationship applies as a first step:

$$P^{rez} \cong P \Leftrightarrow H \cong V^{rez} \Leftrightarrow V \cong H^{rez} \Leftrightarrow x_1 x_4 = x_2 x_3.$$

   The whole chain P is therefore **Symmetrical** if its front and rear subchains are **Reciprocal** to each other.

2. The following conditions are relevant:

$$V \text{ harmonic} \Leftrightarrow x_1 = x_2 x_3 / (2x_3 - x_2),$$
$$H \text{ arithmetic} \Leftrightarrow x_4 = 2x_3 - x_2.$$

3. In the case where P is symmetric – i.e. $P^{rez} \cong P$ – are equivalent:

(continued)

Theorem 3.4 (continued)
V harmonic $\Leftrightarrow$ H arithmetic.

This is equivalent to the magnitudes satisfying the two equations:

$$x_1 = x_2 x_3 / (2x_3 - x_2) \quad \text{and} \quad x_4 = 2x_3 - x_2.$$

4. Conversely, if V is harmonic and H is arithmetic, then P is symmetric, and then the magnitude equations from (2) or (3) hold.

Proof To (1): It is $P^{\text{rez}} \cong P$ exactly if the similarity $x_1 : x_2 \cong x_3 : x_4$ and therefore for the variables the fraction-arithmetic relation

$$x_1 x_4 = x_2 x_3 \Longleftrightarrow x_1 = x_2 x_3 / x_4$$

and exactly then, if this condition is fulfilled, the chains $x_1 : x_2 : x_3$ and $x_2 : x_3 : x_4$ are by definition reciprocal to each other. This is the case, because they have the proportion $x_2 : x_3$ in common.

To (2) and (3): According to the definition of the harmonic mean, the front chain $V = x_1 : x_2 : x_3$ is harmonic exactly if the equation

(a)
$$x_2 = 2x_1 x_3 / (x_1 + x_3)$$

is valid. The equivalent simple conversion to the variable x_1 is as follows

(b)
$$x_1 = x_2 x_3 / (2x_3 - x_2)$$

The posterior chain H is arithmetic if and only if $x_4 - x_3 = x_3 - x_2$, i.e. if

(c)
$$x_4 = 2x_3 - x_2$$

holds. Do we now have the symmetry relation according to part (1),

(d)
$$x_1 = x_2 x_3 / x_4,$$

then we read off: If the relation (c) holds, afterwards the equations (b) and (d) are equivalent, which shows the equivalence in statement (3).

To (4): If equations (b) and (c) hold, so does equation (d), but according to statement (1) this implies the symmetry of P.

This theorem is applied if we want to find a rear arithmetic 3rd proportional (x) and a front harmonic 3rd proportional (y) for a given proportion $a : b$, so that consequently the complete chain $P = y : a : b : x$ is symmetric.

According to the described context, the problem is quickly solved as follows: The arithmetic 3rd proportional results as a trivial calculation with equation (c). Then the matching harmonic 3rd proportional results simply from equation (b) or from equation (d) in the enumeration of the preceding proof.

The next example block 3.8 shows some integer specified proportion-chains

$$P \cong y : a : b : x$$

to a given 'inner proportion' $a : b$ with a back arithmetic 3rd proportional (x) and a front harmonic 3rd proportional (y). The model chord indications are intended (in the non-Ekmelian cases) for purely diatonic temperament.

Example 3.8 Arithmetic-Harmonic 3. Proportional

$a : b$	x	Y	Similar chain P	Chord model
$1 : 2$	3	$2/3$	$2 : 3 : 6 : 9$	$c_0 - g_0 - g_1 - d_2$.
$2 : 3$	4	$3/2$	$3 : 4 : 6 : 8$	$c_0 - f_0 - c_1 - f_2$
$3 : 4$	5	$12/5 = 2,4$	$12 : 15 : 20 : 25$	$c_0 - e_0 - a_0 - cis_1$
$3 : 5$	7	$2\frac{1}{7}$	$15 : 21 : 35 : 49$	Melodious chord
$4 : 5$	6	$10/3 = 3,\overline{3}$	$10 : 12 : 15 : 18$	$c_0 - es_0 - g_0 - b_1$
$8 : 9$	10	$72/10 = 7.2$	$36 : 40 : 45 : 50$	Diatonic tritone cluster $(9 : 10) \oplus (8 : 9) \oplus (9 : 10)\ b_{-1} - c_0 - d_0 - e_0$
		$9 : 10$	11	$90/11 = 8, \overline{18}$
			$90 : 99 : 110 : 121$	Ekmelic tritone cluster $(10 : 11) \oplus (9 : 10) \oplus (10 : 11)$

◀

All these examples impressively document the basic symmetries that develop in the interplay of

$$[\text{arithmetic} - \text{harmonic}] \text{ with } [\text{reciprocal} - \text{mirrored ordered}]$$

subchains, if they are considered in the light of an overall symmetry. In the following part (B) these relations now find their most general platform:

(B) **Symmetries of the mean value proportion chains**

The following theorem describes in general how Babylonian properties change as chains change to their reciprocals.

Here we do not only make use of the usual modern mathematical language, but we also formulate results which did **not** appear in the ancient view -at least not in this way: Both the proportions and the calculated mean values of these proportions and their 'contra' variants show extraordinarily interesting parallels and symmetries: The designations 'contra' thereby become interpretable under other aspects.

Theorem 3.5 (Symmetries of Babylonian Proportion-Chains)

1. **Invariance under Similarity:** Similar chains also have the same Babylonian property, that is, equivalence holds:

   $A = a_0 : a_1 : \ldots : a_n$ is geometric/arithmetic/harmonic
   $\Leftrightarrow$ all chains that are too similar are geometric/arithmetic/harmonic.

   Thus, with a given proportion chain A, all chains of the equivalence class have

   $$\mathcal{M}(A) = \left\{ A' = a_0' : a_2' : \ldots : a_n' \mid a_0' : \ldots : a_n' \cong a_0 : \ldots : a_n \right\}$$

   the same Babylonian medietary property as A, if it has one.

   The same is true for all other mean value chains such as the contra-harmonic, etc.

2. **Reciprocity Symmetries:** The following symmetry relations hold between a proportion chain A and each of its reciprocals A^{rez}:
   (a) A geometric $\Leftrightarrow A^{\text{rez}}$ geometric,
   (b) A arithmetic $\Leftrightarrow A^{\text{rez}}$ harmonic,
   (c) A harmonic $\Leftrightarrow A^{\text{rez}}$ arithmetic.
   But there are also similar relationships for the other medieties:
   (d) A contra-harmonic $\Leftrightarrow A^{\text{rez}}$ contra-arithmetic,
   (e) A contra-arithmetic $\Leftrightarrow A^{\text{rez}}$ contra-harmonic,
   (f) A subhomothetic $\Leftrightarrow A^{\text{rez}}$ contra-superhomothetic.

Some Remarks

1. Thus, reciprocals and inverse orderings convert an arithmetic into a harmonic, a contra-arithmetic into a contra-harmonic chain of proportions, and vice versa in each case.
2. Because the reciprocal of a reciprocal is similar to the initial proportion chain,

$$A \cong (A^{\mathrm{rez}})^{\mathrm{rez}},$$

in the preceding list (2) of symmetries some things are actually mentioned 'superfluously'. Thus (2b) is automatically equivalent to (2c): If we apply the symmetry (2b) to the chain A^{rez}, then it follows that

$$A^{\mathrm{rez}} \text{ arithmetic} \Leftrightarrow (A^{\mathrm{rez}})^{\mathrm{rez}} \text{ harmonic.}$$

Thus, according to the invariance property (1), A is also harmonic and vice versa, so we have obtained the symmetry (2c). We have only listed these things separately for reasons of a more drastic perception.

3. Both from the formulas of Theorem 3.3 and also because of the diametrical positions in the order of medieties described there (also with respect to the geometric mean as well as among each other) and now also from the symmetries of reciprocity in Theorem 3.5, further possibilities can be found as to how the terms of the 'contra-means' to the other medieties can be interpreted.

4. The property 'geometric' is the only one that remains invariant to the reciprocal construction – this already underlines the singular importance of this mediate. In the following Sect. 3.5 (namely in Theorem 3.7) we encounter further symmetry principles of the geometric mean, which underline even more its special role.

Proof of Theorem 3.5

To (1): Because in the case of similarity of two proportion chains all 'inner' proportions (and especially all equally positioned step proportions) are preserved, nothing changes in the defining properties.

Nevertheless we want (as it were for **the exercise of proportional mathematics**) to carry out the details once concretely. However, due to Definition 3.3, we can restrict ourselves to two-stage-chains, and here we consider two cases which stand as models for all others.

Let $a : b : c$ be a proportion chain and let $a' : b' : c'$ be similar to $a : b : c$, symbolically $a : b : c \cong a' : b' : c'$. Then, by definition and thanks to our definition with Theorem 2.2, this means that simultaneously the two mutually equivalent similarity criteria

$$a : b \cong a' : b' \text{ and } b : c \cong b' : c' \text{ (Stage criterion)}$$
$$a : a' \cong b : b' \cong c : c' \text{(Magnitude criterion)}$$

are available and therefore all these proportions are present. In particular, it follows from the second criterion – by applying the difference-rule – that the differences are also in the same proportion, that is to say:

$$(b-a) : (b'-a') \cong (c-b) : (c'-b') \cong a : a' \cong b : b' \cong c : c'.$$

And from this we also obtain the proportions with the help of the cross-rule

$$(b-a) : a \cong (b'-a') : a' \text{ and } (c-b) : c \cong (c'-b') : c'.$$

Let us now first consider the case where $a : b : c$ is geometric. Thus, the square-rectangle formula in proportion form

$$a : b = b : c$$

is valid, then we recognize directly with the step criterion the equalities

$$a' : b' \cong a : b \cong b : c \cong b' : c'.$$

So also $a' : b' \cong b' : c'$, and the proportion chain a' : b' : c' is geometric.

That was very easy. However, things are a little more subtle for all other cases: Let's next take the case of a harmonic-chain $a : b : c$. It is harmonic exactly when, in the 'piece to the rest' form, the similarity of proportions is

$$(b-a) : (c-b) \cong a : c$$

exists. This is via cross-rule equivalent to the proportion similarity

$$(b-a) : a \cong (c-b) : c,$$

this is the 'piece to piece' form. Now we put everything together:

$$(b'-a') : a' \cong (b-a) : a \cong (c-b) : c \cong (c'-b') : c'.$$

But then the resulting relation is

$$(b'-a') : a' \cong (c'-b') : c'$$

exactly the 'piece to piece' form of a harmonic proportion, which, with the cross-rule, changes back into the familiar 'piece to rest' form.

All other cases (arithmetic, contra-arithmetic and contra-harmonic) are copies of this (certainly tricky) procedure.

To (2): Considering the formally 'break-free' form $ab : ac : bc$ for a reciprocally similar proportion chain to $a : b : c$ all statements (justified by the invariance properties (1)) can be done in a procedure following the proof to (1). We again consider two cases by way of example:

1. Case: The chain $a : b : c$ is geometrically $\Leftrightarrow a : b \cong b : c$. Then immediately follow for the reciprocal $ab : ac : bc$ the relations

$$ab : ac \cong b : c \ \text{ and } \ ac : bc \cong a : b,$$

from which then also the similarity

$$ab : ac \cong ac : bc,$$

so that the chain $ab : ac : bc$ is indeed geometric, since it satisfies the square-rectangle formula.

2. Case: The chain $a : b : c$ is arithmetic: Then $(c - b) : (b - a) \cong b : b$. Now we apply the inverse of the shortening-rule twice and the inverse-rule once, then follows:

$$a(c - b) : (b - a) \cong ab : b \Leftrightarrow (ac - ab) : (b - a) \cong ab : b$$
$$\Leftrightarrow (b - a) : (ac - ab) \cong b : ab \Leftrightarrow c(b - a) : (ac - ab) \cong cb : ab$$
$$\Leftrightarrow (cb - ca) : (ac - ab) \cong cb : ab \Leftrightarrow (ac - ab) : (cb - ca) \cong ab : bc.$$

However, the last condition, is precisely the one on the basis of which the proportion chain

$$ab : ac : bc$$

is harmonic by definition.

The remaining cases are analogous to this case and may perhaps serve as practice for the readers.

Nevertheless, we want to show what the **functional equation of** the **mean function of** Theorem 3.1 is capable of, and prove by way of example that the subhomothetic and the contra-superhomothetic proportion chains are reciprocal to each other.

Again, as before, it suffices to do this for 2-step proportion chains. In order to apply this wonder weapon, we simply consider the corresponding division parameters – these are the respective values of the **mean-value-function** $f(x)$ and we read these from the table of Theorem 3.3:

$$x_{VIII} - \text{subhomothetic} \Rightarrow f(x_{VIII}) = a/(b-a),$$
$$x_{XI} - \text{contra-superhomothetic} \Rightarrow f(x_{XI}) = (b-a)/b.$$

Then statement (5) of Theorem 3.1 provides an ingenious criterion, which in verbal formulation reads thus:

- **Hyperbola Criterion:** Exactly if the product of two division parameters is equal to a/b, then the product of the associated medietes is equal to a*b.

Now, in our present case, in fact

$$f(x_{VIII}) * f(x_{XI}) = a/(b-a) * (b-a)/b = a/b.$$

Therefore, the hyperbolic equation $x_{VIII}{}^* x_{XI} = ab$, applies, which means nothing else than that the similarities (duplicated and mutually equivalent due to the cross-rule) are

$$a : x_{VIII} \cong x_{XI} : b \text{ and } a : x_{XI} \cong x_{VIII} : b$$

and that is why the two chains

$$a : x_{VIII} : b \text{ and } a : x_{XI} : b$$

reciprocal to each other, because their step proportions are similar when interchanged.

We describe this interplay between magnitudes and their division parameters again at the end of the following Sect. 3.5, and give it the name of a 'magic formula'.

Question What is the situation for the medietary pair x_{IX} (contra-subhomothetic) and x_X (superhomothetic)?

3.5   Geometrical Means: The Power Centre of the Medieties

The geometric mean of two magnitudes $a < b$ obviously has a clear disadvantage compared to the other means (we mean the arithmetic mean, harmonic mean, their contra-forms, but also the means called 'homothetic'): One needs the square root calculation. For all the others only simplest fractional arithmetic calculations are needed – mental arithmetic is sufficient, so to speak and independent of this arithmetic, which is regarded as 'superior', there is the aggravating fact, that the root of the product of two given numbers is rational rather by chance, but as a rule non-rational (irrational) and therefore in the sense of ancient notions of numbers was called 'not specifiable' or also 'not nameable', but this does not mean that the 'existence of this root' would have been unknown to the ancient teachers all around.

About geometric-chains of proportions one finds in Euclid *(Euclid, Theorem 11 Book VIII)* the observations:

1. *There is always a geometric mean between two square numbers.*
2. *For two square numbers there is always a number, which represents the geometric mean of the two square numbers or which forms a geometric proportion with them.*

So if a and b are themselves natural numbers, then with $c = ab$ is also $a^2 : c \cong c : b^2$. And thus with $A = a^2 : ab : b^2$ a geometric chain of proportions is always found.

3. *If n is a natural number, then there is a (integer) geometric proportion chain a : n : b exactly when n is a product of two different integer factors.*

Thus, for an integer n, the chain $a : n : b$ is a geometric proportion chain exactly when $n = pq$ is a product with integer factors $p \neq q$.

4. *If a number arises from the fact that two 'similar' numbers multiply each other, then the product must be a square number. (Euclid, Book IX).*
   If a, b, c, d are natural numbers with $a : b \cong c : d$, then the product of all numbers $(a * b * c * d)$ is a square number. (Theorem 1, Book IX)

As a result in the ancient musical chains of proportions we seem almost never to encounter the geometric mean directly. The field is apparently left exclusively to the arithmetic mean, its harmonic partner, and even their two contra-medieties.

So Does the Geometric Mean not Make an Appearance?

Not at all! Already in the game of front and back third proportionals we met equations (or the considerations accompanying them), which showed us the connection to the hyperbola of Archytas as fast as lightning and everywhere. Thus, in Theorem 3.4 the equation told us

$$x_1 x_4 = x_2 x_3,$$

that the opposite pairs of magnitudes (x_1, x_4) and (x_2, x_3) of the proportion chain $P = x_1 : x_2 : x_3 : x_4$ lie on the same hyperbola of Archytas. However, this is the unique hyperbola, which passes through its point of symmetry (z, z), and then z is simultaneously the geometric mean of both pairs of magnitudes

$$z = \sqrt{x_1 x_4} = \sqrt{x_2 x_3}.$$

In other words:

*In the background, one quantity, **the geometric mediety,** controls all things that lie within the haze of symmetrical constellations.*

How great this power really is, we can see for the time being in the following theorem, which summarizes the most important things about it, but only the following Chap. 4, in which we will examine nested mediate sequences, will really demonstrate how this geometric mediate controls things.

Theorem 3.6 (Symmetry Principles of Geometric Mediety)
The following symmetry criteria for proportion chains are equivalent. However, they describe the controlling influence of geometric mediety in different ways.

1. **Principle of Symmetry Center**
   Let $A = a_0 : a_1 : \ldots : a_n$ be a proportion chain, and let $z_{\text{geom}}(a_0, a_n)$ be the geometric mean of the total proportion of A. Then holds:

   $$A \cong A^{\text{rez}} \Leftrightarrow a_k * a_k^* = a_k * a_{n-k} = \left(z_{\text{geom}}\right)^2 \quad \text{for all } 0 \leq k \leq n$$
   $$\Leftrightarrow z_{\text{geom}}(a_0, a_n) = z_{\text{geom}}(a_k, a_{n-k}) \quad \text{for all } 0 \leq k \leq n.$$

   Conclusion: The geometric mean z_{geom} of the total proportion is therefore in the case of symmetry of the proportion-chain simultaneously the geometric mean of all pairs of magnitudes mirrored to each other – thus z_{geom} is the **center of symmetry of** A. Conversely, the chain is symmetrical, which provides that the geometric mean of the total proportion $z_{\text{geom}}(a_0, a_n)$ is the center of symmetry of the whole chain.

2. **The Symmetry Principle of the Hyperbola of Archytas**
   Let m ascending 2-step proportion chains be given

   $$A_1 = a_1 : z : b_1, \ldots, A_m = a_m : z : b_m,$$

   which therefore all have the same mean proportional z. Then the following statements are equivalent:
   (a) All pairs of magnitudes $(a_1, b_1), \ldots, (a_m, b_m)$ lie on the hyperbola of Archytas $xy = z^2$, which passes through the point (z, z) as the point of symmetry.

(continued)

Theorem 3.6 (continued)
(b) All chains $A_1, \ldots, A_m$ are geometric proportion chains: The common magnitude z is thus geometric mean simultaneously for all pairs of magnitudes $(a_1, b_1), \ldots, (a_m, b_m)$, which can also be expressed in this way:

$$z = z_{\text{geom}}(a_1, b_1) = z_{\text{geom}}(a_2, b_2) = \ldots = z_{\text{geom}}(a_m, b_m).$$

(c) The $2m$- or $(2m + 1)$-step proportion chain, which is constructed mirror-symmetrically as follows

$$P = a_1 : a_2 : \ldots : a_m : (z :) b_m : \ldots : b_2 : b_1$$

is symmetric, that is, $P \cong P^{\text{rez}}$.

Remark For the definition of this proportion chain P it is not important to arrange the magnitudes in ascending order a_k – only the contra-positioned enumeration of the involved pairs of magnitudes is important. However, if the numbering is set up in such a way that the magnitudes already have the ascending order

$$a_1 \leq a_2 \leq \ldots \leq a_m,$$

then in the case of symmetry of P, respectively in the case that the magnitude z is a common geometric mean, also the partner magnitudes b_k of a_k are ordered (inversely),

$$b_m \leq \ldots \leq b_2 \leq b_1,$$

and one obtains the whole chain P as an ascending proportion chain of the whole magnitude sequence. It can be regarded as a **nested construction** from the proportions $A_1, \ldots, A_m$. (We have added the equality case more or less for reasons of as much generality as possible).

Proof to (1) If $A = a_0 : a_1 : \ldots : a_n$ is symmetric, then for an **even number of steps** with the representation $n = 2m$, it holds that there is a mean magnitude, and this is then a_m; it is then also the geometric mean of the total proportion

$$z_{\mathrm{geom}}(a_0, a_n) = a_m,$$

And therefore a_m is also the geometric mean of all magnitudes mirrored at position m

$$a_m = z_{\mathrm{geom}}(a_{m-k}, a_{m+k}), \quad k = 1, \ldots, m.$$

This means nothing else than that all (m) 2-step subchains of A of the form

$$a_{m-j} : a_m : a_{m+j}, \quad j = 1, \ldots, m$$

are geometric. A short justification for this would be: Because A is symmetric (yes, by the mirror principle of Theorem 2.5), it holds specifically that the similarities are

$$a_j : a_m \cong a_m^* : a_j^* = a_{n-m} : a_{n-j} \cong a_m : a_{2m-j} \text{ for all } j = 0, \ldots, n$$

are fulfilled. So with the index renaming $k = m - j$ the geometric proportions arise

$$a_{m-k} : a_m \cong a_m : a_{m+k}.$$

For an **odd number of steps** $n = 2m + 1$ the common geometric mean $z_{\mathrm{geom}}(a_0, a_n)$ of all mirrored magnitudes lies between the middle members a_m and a_{m+1}, whose geometric mean z_{geom} is also, and all ($m + 1$) 2-step subchains of the form

$$a_{m-k} : z_{\mathrm{geom}}(a_0, a_n) : a_{m+1+k}, k = 0, \ldots, m$$

are geometric.

The derivation is similar to the previous one. Certainly, we can also use the hyperbola principle of Theorem 2.5: According to this, all products $(a_k * a_k^*)$ of the mirrored pairs of magnitudes are equal if and only if A is symmetric:

$$a_0 * a_n = a_1 * a_{n-1} = \ldots = a_n * a_0 = \left(z_{\mathrm{geom}}(a_0, a_n)\right)^2,$$

and then the proportion equations follow from this

$$a_k : z_{\mathrm{geom}}(a_0, a_n) \cong z_{\mathrm{geom}}(a_0, a_n) * a_k^*,$$

which is equivalent to our assertion. The subsequent positional description of the geometric mean results from the symmetry of arrangement of the mirrored magnitudes: In the even case namely

$$a_m = a_{2m-m} = a_m^*,$$

and in the odd case

$$a_m^* = a_{2m+1-m} = a_{m+1}.$$

Proof of (2): The hyperbola principle of the Theorem (2.5) is equivalent to the symmetry of P and from it follows everything desired, but also another way would lead to the goal: If P is symmetric, this is equivalent to all pairs of magnitudes mirrored to each other having similar proportions. So in our enumeration pattern this leads to the relations

$$a_0 : a_k \cong b_k : b_0 \text{ for all } k = 1, \dots, n.$$

This in turn means the equality of all products

$$a_0 * b_0 = a_k * b_k \text{ for all } k = 1, \dots, n.$$

Accordingly, all proportions A_k have the same center of symmetry (z), which is the common geometric mediety. Thus, the theorem is shown.

To conclude this discussion of geometric means with their widely interconnected sphere of action, we come perhaps to the most impressive aspect, *which also connects geometry and calculus with the proportion theory of medieties.*

This is a result which has its origin in the functional equation of the mean function of Theorem 3.1, and which we now present in a very memorable application. For this purpose we want to call very briefly again the proportion function f:

For any intermediate value $a < x < b$ the function

$$y = f(x) = (x - a)/(b - x)$$

has the task of specifying the division ratio – or better: The division parameter (y), of the two segments of the total distance $(b - a)$ conditioned by the given point x to each other. We have already discussed this function in great detail in the first Sect. 3.1 of this chapter and in Theorem 3.3 the mean formulas as well as the division parameters are listed. For the geometric mean we have there the form

$$f(z_{\text{geom}}) = \sqrt{a/b} = q^{-1/2}$$

with $q = b/a$ found. Now, we recall the functional equation of Theorem 3.1, according to which for two magnitudes x_1, x_2 with $a < x_1 < b$ and $a < x_2 < b$ the equivalence

$$x_1 * x_2 = ab \Leftrightarrow f(x_1) * f(x_2) = a/b$$

exists. The left side says that the two magnitudes lie on the hyperbola of the archytas. Thus, also have z_{geom} as geometric mean, but the right side says that the two division parameters $f(x_1), f(x_2)$ of the magnitudes x_1, x_2 have the division parameter $f(z_{\text{geom}})$ as geometric mean, because we have the equation

$$f(x_1) * f(x_2) = \frac{a}{b} = \left(f\left(z_{\text{geom}}\right)\right)^2$$

won. Thus, the proportion chain

$$f(x_1) : f\left(z_{\text{geom}}\right) : f(x_2)$$

is a geometrical chain and therefore also a symmetrical one. We write down this result, which can be interpreted in this way, in the last Theorem 3.7 of this section. It rounds off in a drastic way our researches into inner connections of medieties, their proportions and symmetries.

Theorem 3.7 (The Magic Formula of Geometric Means: The Harmonia Perfecta Maxima Abstracta)

For any two magnitudes x_1, x_2 with $a < x_1 < b$ and $a < x_2 < b$ equivalence holds:

$$x_1 * x_2 = z_{\text{geom}} * z_{\text{geom}} = ab \Leftrightarrow f(x_1) * f(x_2) = f(z_{\text{geom}}) * f(z_{\text{geom}}) = a/b.$$

Thus, the division parameters $f(x_1), f(x_2)$ can be used to read off one-to-one the symmetries of the proportions with respect to the data x_1, x_2: This result has the following list of equivalent interpretations:

(A) The left side of the equivalence describes the relations of the two **Magnitudes** x_1, x_2 to the geometric mean z_{geom}. This condition

$$[x_1 * x_2 = z_{\text{geom}} * z_{\text{geom}} = ab]$$

has the following mutually equivalent interpretations:
1. The numerical value of the product of the means $x_1 * x_2 = ab$.
2. The similarity proportions apply

(continued)

Theorem 3.7 (continued)
$$a : x_1 \cong x_2 : b \text{ respectively } a : x_2 \cong x_1 : b.$$

3. The geometric mean of (a, b) is also geometric mean of (x_1, x_2),

$$z_{\text{geom}}(a, b) = z_{\text{geom}}(x_1, x_2),$$

and that in turn means the similarity

$$x_1 : z_{\text{geom}} \cong z_{\text{geom}} : x_2.$$

4. The proportion chain $P = x_1 : z_{\text{geom}} : x_2$ is geometric.
5. The proportion chain $P = x_1 : z_{\text{geom}} : x_2$ is symmetrical, i.e. $P \cong P^{\text{rez}}$.
6. The **Abstract Musical Canon** $A_{\text{mus}} = a : x_1 : x_2 : b$ is symmetrical: $A_{\text{mus}} \cong (A_{\text{mus}})^{\text{rez}}$.
7. The point (x_1, x_2) lies on the **Hyperbola of Archytas** $y * x = ab$, which runs through the points (a, b) *and* (b, a) and has its point of symmetry as an intersection with the bisector $y = x$ in the point $\left(\sqrt{ab}, \sqrt{ab}\right)$, i.e. in the geometric mean $(z_{\text{geom}}, z_{\text{geom}})$.

(B) The right-hand side of the equivalence, on the other hand, describes the relations of the two *division parameters* $f(x_1)$, $f(x_2)$ to the division parameter of the geometric mean. Overall, the condition

$$f(x_1) * f(x_2) = \left(f\left(z_{\text{geom}}\right)\right)^2 = a/b$$

has the following mutually equivalent interpretations:
1. The numerical value of the product of the division parameters is $f(x_1) * f(x_2) = a/b$.
2. The similarity proportion applies

$$f(x_1) : f\left(z_{\text{geom}}\right) \cong f\left(z_{\text{geom}}\right) : f(x_2).$$

3. The proportion chain of the **division parameters**

$$T = f(x_1) : f\left(z_{\text{geom}}\right) : f(x_2) = f(x_1) : \sqrt{a/b} : f(x_2)$$

is geometric.

(continued)

Theorem 3.7 (continued)
4. The proportion chain of the division parameters

$$T = f(x_1) : \sqrt{a/b} : f(x_2)$$

is symmetrical. Thus, reciprocal to itself, i.e. $T \cong T^{\text{rez}}$.
5. The point $(f(x_1), f(x_2))$ lies on the hyperbola $y * x = a/b$, which has its point of symmetry as an intersection with the bisector $y = x$ at the point $\left(\sqrt{a/b}, \sqrt{a/b}\right)$, i.e. in the division parameters of the geometric mean

$$\left(f\left(z_{\text{geom}}\right), f\left(z_{\text{geom}}\right)\right)$$

to the division parameters $f(x_1)$, $f(x_2)$.

Conclusion: By combining all possible equivalent forms of statement group (A) with those of statement group (B), an imposing network of equivalent descriptions around the symmetry of two general averages in terms of their relation to the geometric mediety emerges. Summing up, we find in the following short form all the foregoing in a handy formula, so to speak in a **magic formula** of geometric medieties in theory and practice, which describes the symmetry of a most general possible canon:

Harmonia perfecta maxima abstracta:
With the proportion chains of two medietes x_1, x_2 and their division parameters

$$A_{\text{mus}} = a : x_1 : x_2 : b \quad \text{and} \quad T = f(x_1) : \sqrt{a/b} : f(x_2)$$

we have the theoretical and computationally practical symmetry relation:

$$\text{concerning the theory} : A_{\text{mus}} \cong A_{\text{mus}}^{\text{rez}} \Leftrightarrow T \cong T^{\text{rez}}$$
$$\text{concerning the practice} : x_1 * x_2 = ab \Leftrightarrow f(x_1) * f(x_2) = a/b.$$

The first form describes all the inner proportional relationships of the **means of the magnitudes** a, b, the second gives the numerical criteria to these symmetries by means of the **Division Parameters.** Both forms are equivalent and express a single abstract-musical principle.

It is precisely on this principle that we gave the proof at the end of Sect. 3.4 that the reciprocals of subhomothetic proportion chains are contra-superhomothetic and for the pleasure of these amazingly short proofs we test the symmetry of a proportion chain A also stated in Theorem 3.5 there,

$$A \text{ contra-arithmetic} \Leftrightarrow A^{\text{rez}} \text{ contra-harmonic,}$$

with precisely this principle of harmonia perfecta maxima abstracta:

If x_1 is the contra-arithmetic and if x_2 is the contra-harmonic mediety of a, b, then the division parameters are according to the statements in Theorem 3.3

$$f(x_1) = a^2/b^2 \ and \ f(x_2) = b/a.$$

Consequently, their product fulfils the equation

$$f(x_1) * f(x_2) = a/b,$$

and the assertion is shown, because as a result, according to this principle, the two mediævalities lie mirrored to the geometrical mediævality and hence mirrored to the edges a, b, but this means that the chains of proportions

$$a : x_1 : b \ \text{and} \ a : x_2 : b$$

are reciprocal to each other. Totally simple, isn't it?

▶   *Note again that in this procedure we do not even need to know the medieties explicitly as formulas – merely knowing their division parameters is sufficient. ('Tell me your values, and I'll tell you who you are').*
   In praise of theory.

3.6   The Harmonia Perfecta Maxima Diatonica

In the previous sections, we have frequently encountered musical chains of proportions in the form of the 'canon' and many things have already been reported about them. In particular, we have illuminated the pioneering aspect (from the point of view of music-theory) of stepped Babylonian chains of proportions in terms of their inner proportions, symmetries and other regularities. This was the subject of Theorem 3.2, Harmonia perfecta maxima babylonica, for the special case of the series of proportions (magnitude a) : (harmonic mediety) : (arithmetic mediety) : (magnitude b).

Moreover, in the preceding Sect. 3.5 we have just presented this 3-level canon in its most general form possible by means of Theorem 3.7 as 'Harmonia perfecta maxima abstracta'.

Nevertheless, in order to introduce a further central theorem of music theory, we would like to take up the historical-musical idea of the Harmonia perfecta maxima once again and place it before the further development of the Harmonia.

Therefore, from time immemorial, the series of numbers

$$6 - 8 - 9 - 12$$

was considered as the epitome of a fusion of arithmetic-harmonic symmetries and their proportions with musical elements. The numerical series of this canon embodies the structure of an octave 6 : 12 as a construct (namely the sum, see section 5.1) of three Pythagorean intervals

Fourth $\oplus$ Tonos $\oplus$ Fourth.

In addition, the intermediate magnitudes decisive for this are the harmonic mean (8) and the arithmetic mean (9) of the octave magnitudes 6 and 12. Fascinating for the science of that time and proving its supernatural character, were above all the remarkable symmetries of proportions of this 4-membered chain: Thus the following observations are immediately before our eyes:

1. $6 : 8 \cong 9 : 12$ (fourths).
2. $6 : 9 \cong 8 : 12$ (fifths).
3. The proportion chains $6 : 9 : 12$ and $6 : 8 : 12$ are reciprocal to each other. The former is arithmetic and the latter is harmonic.
4. The proportion chains $6 : 8 : 9$ and $8 : 9 : 12$ are also reciprocal to each other and represent the forms of the fifth structure of fourths and tonos.

The decomposition of the octave into the two forms

$$6 : 9 : 12 \ \cong \ 2 : 3 : 4 \text{ and } 6 : 8 : 12 \ \cong \ 3 : 4 : 6$$

corresponds to the interval divisions

Octave $-$ chain $(6 : 9 : 12) =$ Fifth $\oplus$ Fourth (arithmetic $-$ or also authentic decomposition),
Octave $-$ chain $(6 : 8 : 12) =$ Fourth $\oplus$ Fifth (harmonic $-$ or also plagalic decomposition).

This concept of the division of an interval according to an arithmetic or harmonic proportion structure, incidentally, ultimately originates from the concept of the opposing proportion series of the Perissos and Artios numbers meeting in the unit 1 : 1, as we indicated in Sect. 1.1:

> The authentic octave is the octave divided according to an arithmetic chain of proportions, and the plagalic octave is the same if one follows the downward direction and then switches from the perissos to the artios magnitudes:

$$2 : 3 : 4 (\text{upward octave}) \rightleftarrows (1/2) : (1/3) : (1/4) \cong 6 : 4 : 3 \ \ (\text{downward octave})$$

Then the interval sequences are also the same: first the fifth, then the fourth. In the first case upwards, in the second case downwards. (see [6], 1st chap.)

The symmetry of the structure of the fifth as *Fourth* $\oplus$ Whole tone or as *Whole tone* $\oplus$ *Fourth* shown in the preceding point (4). On the other hand, is not of an authentic-plagal nature. Such divisions would be the proportion chains

$$12 : 15 : 18 \cong 60 : 75 : 90 \ \ (\text{arithmetic} - \text{authentic}),$$
$$10 : 12 : 15 \cong 60 : 72 : 90 \ \ (\text{harmonic} - \text{plagalic}),$$

whose difference in the mean value proportion

$$72 : 75 \cong 24 : 25,$$

corresponds to a so-called **'minor or small chroma'**, the difference of the diatonic semitone 15 : 16 in the minor whole tone 9 : 10.

However, this (and actually almost all significant intervals of the diatonic) cannot be derived from the **prime numbers 2 and 3** alone. Thus, also not from the Pythagorean canon. Here, the extension of the arithmetic and harmonic means by their contra-means produces a wealth of new symmetries as well as new interval constructions. Applied to the Pythagorean canon, we thus arrive at the number chains (the **complete diatonic canon**)

$$6 - 7.2 - 8 - 9 - 10 - 12 \ \ \text{respectively} \ \ 30 - 36 - 40 - 45 - 50 - 60,$$

for which an enumeration of all inner symmetries of many of its partial proportions, together with their musical meanings, opens up a marvelous world of its own. It is therefore easy to understand why the attributes 'perfecta' (perfect) and 'maxima' (greatest perfection) were added to this mathematical-musical miniature and why it was chosen as the **Universal Law of Music-Theory.**

It remains to be noted that the expansion of the simple octave canon resulted not least from the demand (highly controversial at the time) for the **addition of the pure major**

third $4 : 5 \cong 8 : 10$ to the existing Pythagorean set of intervals. In any case, this demand turned the Pythagorean canon into the diatonic (octave) canon and the **temperament age** that began in the late Middle Ages, with its heyday around the era of Johann Sebastian Bach, has much to tell us about this.

▶   **Important**

Now, in the following theorem we show that the beauty of this structure does not depend solely on the smooth octave numbers of the canon: All symmetries are explained exclusively by the mean character of their magnitudes and we can by the joint force of our main results, viz.

- the Theorem 2.5 about the symmetry criteria,
- the Proportion Chain Theorem 2.6,
- the Theorem 3.4 about the symmetry of the 3rd proportional
- and, above all, the reciprocity rules of Theorem 3.6

gain our **main result of ancient diatonics** in the proportion form.

Theorem 3.8 (The Harmonia Perfecta Maxima of the Pure Diatonic)

Let $A = a : b$ be any proportion such that the musical interval $[a, b]$ is ascending. Then the central statement is:

Harmonia perfecta maxima diatonica: The chain divided by the four musical medieties of A into a 5-step proportion chain

$$D_{\mathrm{mus}}(a, b) = a : x_{\text{co-arith}} : y_{\mathrm{harm}} : x_{\mathrm{arith}} : y_{\text{co-harm}} : b$$

is a symmetrical, ascending proportion chain: $D_{\mathbf{mus}} \cong D_{\mathbf{mus}}^{\mathrm{rez}}$.

From this symmetry and from the mean value equations there follows in turn a considerable list of internal similarities of proportions which include, among other things, an abstract version of the theorems of Iamblichos and Nicomachus such as these:

1. **Center symmetry – Theorem of Nicomachus:** The geometric mean is a common mean of diametrically positioned magnitudes: The chain D_{mus} always has at least the three geometric 2-step subchains

---

(continued)

Theorem 3.8 (continued)

$$G_1 = a : z_{\text{geom}} : b,$$

$$G_2 = y_{\text{harm}} : z_{\text{geom}} : x_{\text{arith}},$$

$$G_3 = y_{\text{co-arith}} : z_{\text{geom}} : x_{\text{co-harm}}.$$

2. **Babylonian Partial-Chain Structure – Theorem of Iamblichos:** The proportion chain D_{mus} always contains at least two arithmetic 2-step partial chains,

$$A_1 = a : x_{\text{arith}} : b,$$

$$A_2 = y_{\text{harm}} : x_{\text{arith}} : y_{\text{co-harm}},$$

and consequently also always two harmonic 2-step partial chains,

$$H_1 = a : x_{\text{harm}} : b,$$

$$H_2 = x_{\text{co-arith}} : y_{\text{harm}} : x_{\text{arith}},$$

which are reciprocal to each other in pairs and which at the same time appear as mirrored partial chains of $D_{\textbf{mus}}$, in detail:

$$H_1 = A_1^* \cong A_1^{rez} \quad \text{and} \quad H_2 = A_2^* \cong A_2^{rez}.$$

The symmetrically arranged 3-step partial chain of all musical medieties

$$M_{\text{mus}} = x_{\text{co-arith}} : y_{\text{harm}} : x_{\text{arith}} : y_{\text{co-harm}}$$

is itself again a symmetrical proportion chain.
3. In addition, there is a relationship of mutual equivalence between them:
   Harmonia perfecta maxima diatonica
   ⇔ **Nicomachus' theorem (in the form (1))**
   ⇔ **Theorem of Iamblichos (in the form (2)).**

Proof: The proof of the symmetry $D_{\text{mus}} \cong D_{\text{mus}}^{\text{rez}}$ is given when we have shown the center symmetry, for example, and then use Theorem 3.6. For this we calculate with the help of the mean value formulas the products

$$y_{\text{harm}} * x_{\text{arith}} = \frac{2ab}{a+b} * \frac{1}{2}(a+b) = ab = z_{\text{geom}}^2$$

$$x_{\text{co-arith}} * y_{\text{co-harm}} = \frac{ab}{a^2+b^2} * (a+b) * \frac{a^2+b^2}{a+b} - ab = z_{\text{geom}}^2.$$

Therefore the geometric mean z_{geom} is simultaneously the geometric mean of y_{harm} and x_{arith}, as well as of $x_{\text{co-arith}}$ and $y_{\text{co-harm}}$. And this is exactly the raison why the complete chain D_{mus} is also symmetric according to Theorem 3.6.

Proof for the statement (2): That the chain $A_1 = a : x_{\text{arith}} : b$ is arithmetic is obvious and that the harmonic-chain $H_1 = a : y_{\text{harm}} : b$ is a reciprocal of this, we have seen in Theorem 3.2 with Iamblichos' theorem for the Pythagorean proportion chain

$$a : y_{\text{harm}} : x_{\text{arith}} : b$$

already shown. Let us now look at the proportion chain

$$A_2 = y_{\text{harm}} : x_{\text{arith}} : y_{\text{co-harm}}.$$

That A_2 is also arithmetic can be seen (even without calculation) from the fact that the harmonic and the contra-harmonic mean divide the distance $(b - a)$ in the opposite proportions $a : b$ and $b : a$, respectively: Thus,

$$y_{\text{harm}} - a = \frac{a}{b+a}(b-a) \text{ and } y_{\text{co-harm}} - a = \frac{b}{a+b}(b-a),$$

and with $x_{\text{arith}} - a = \frac{1}{2}(b-a)$ follows the equality of the differences

$$x_{\text{arith}} - y_{\text{harm}} = \frac{1}{2(b+a)}(b-a)^2 = y_{\text{co-harm}} - x_{\text{arith}}.$$

In the proportion chain

$$H_2 = x_{\text{co-arith}} : y_{\text{harm}} : x_{\text{arith}}$$

now, due to the symmetry of D_{mus}, the front proportion $x_{\text{co-arith}} : y_{\text{harm}}$ is similar to the back proportion of A_2,

$$x_{\text{co-arith}} : y_{\text{harm}} \cong x_{\text{arith}} : y_{\text{co-harm}},$$

and the posterior proportion $y_{\text{harm}} : x_{\text{arith}}$ of H_2 is anyway identical to the anterior proportion of A_2. Therefore H_2 is reciprocal to A_2, and H_2 is a subproportion-chain of D_{mus}. The connection to the mirrored partial chain beyond this is justified by the proportion chain theorem: Subchains which are reciprocal to a subchain are simultaneously also proportion chains of the mirrored magnitudes (taking into account the composition which is also mirrored). Thus, the theorem is finally explained.

We now present three examples:

Example 3.9 The Diatonic Octave canon D_{mus} (1 : 2)

This musical proportion-chain is traditionally incorporated into the outer octave proportion with magnitudes 6:12, and then the most famous proportion-chain of the pure diatonic is created

$$D_{\text{mus}} = D_{\text{mus}}(6 : 12) = 6 : 7,2 : 8 : 9 : 10 : 12 \cong 30 : 36 : 40 : 45 : 50 : 60.$$

In addition to its proportion chain symmetry $D_{\text{mus}} \cong D_{\text{mus}}^{\text{rez}}$ it possesses the additional property of having, first of all, in sum, exactly four arithmetic 2-step partial-chains A_1, ..., A_4, and then there is also the fact that their harmonic reciprocals $H_1, \ldots, H_4$ are also partial-chains of D_{mus}. This gives rise to simultaneously authentic and plagal decompositions of four intervals. We list all these four pairs of chains:

$$A_1 = a : x_{\text{arith}} : b = 6 : 9 : 12$$
$$H_1 = a : y_{\text{harm}} : b = 6 : 8 : 12.$$

This proportion chain pair defines the chordal decomposition of the octave into the two structural forms

Octave $-$ chain $(6 : 9 : 12) = $ Fifth $(2 : 3) \oplus $ Fourth $(3 : 4)$ (arithmetic $-$ authentic)
Octave $-$ chain $(6 : 8 : 12) = $ Fourth $(3 : 4)$ $\oplus$ Fifth $(2 : 3)$ (harmonic $-$ plagalic).

The next proportion chain pair is

$$A_2 = y_{\text{harm}} : x_{\text{arith}} : y_{\text{co-harm}} \cong 8 : 9 : 10, H_2 = x_{\text{co-arith}} : y_{\text{harm}} : x_{\text{arith}} \cong 7,2 : 8 : 9.$$

Here we have the chordal decomposition of the major pure third $8 : 10 \cong 4 : 5$ in the steps of two whole tones:

Third $-$ chain $(72 : 81 : 90) = $ Tonos $(8 : 9) \oplus $ diat.whole tone $(9 : 10)$ (arithmetic $-$ authentic)
Third $-$ chain $(72 : 80 : 90) = $ diat.whole tone $(9 : 10) \oplus $ Tonos $(8 : 9)$ (harmonic $-$ plagalic).

While, according to our Theorem 3.8 above, these two pairs of proportion chains can be found in their general structure in any musical canon with arbitrary dates a and b, in this case of the octave there are two more pairs, viz.

$$A_3 = a : y_{\text{harm}} : y_{co\text{-harm}} = 6 : 8 : 10$$
$$H_3 = x_{\text{co-arith}} : x_{\text{arith}} : b = 7,2 : 9 : 12.$$

The outer proportion $6 : 10$ forms a pure major sixth $3 : 5$, whose decomposition is then as follows

Major Sixth $-$ chain $(12 : 16 : 20) = $ Fourth $(3 : 4) \oplus$ Third $(4 : 5)$ (arithmetic $-$ authentic)
Major Sixth $-$ chain $(12 : 15 : 20) = $ Third $(4 : 5) \oplus$ Fourth $(3 : 4)$ (harmonic $-$ plagalic).

and we find familiar musical realizations again: authentically, this is a 'fourth sixth chord'. For example, $c - f - a$, while plagalically, it is a minor chord in the third position, for example, $c - e - a$.

Finally, we find a fourth pair of arithmetic-harmonic subchains:

$$A_4 = y_{\text{harm}} : y_{\text{co-harm}} : b = 8 : 10 : 12$$
$$H_4 = a : x_{\text{co-arith}} : x_{\text{arith}} = 6 : 7.2 : 9.$$

This is the decomposition of a fifth $8 : 12 \cong 6 : 9 \cong 2 : 3$ into the familiar major and minor triadic forms, viz.

Fifth $-$ chain $(20 : 25 : 30) = $ Major Third $(4 : 5) \oplus$ Minor Third $(5 : 6)$ (arithmetic $-$ authentic)
Fifth $-$ chain $(20 : 24 : 30) = $ Minor Third $(5 : 6) \oplus$ Major Third $(4 : 5)$ (harmonic $-$ plagalic).

The often cited musical realization would be, for example, $c - e - g$ (major, authentic) or $a - c - e'$ (upward, A minor, plagalic).

Besides these surprisingly many arithmetic 2-step partial chains, there is even exactly one 3-step arithmetic partial chain A_0 in D_{diat} (also consequently), whose 3-step harmonic reciprocal $H_0 \cong A_0^{\text{rez}}$ is also included in the canon:

$$A_0 = a : y_{\text{harm}} : y_{\text{co-harm}} : b = 6 : 8 : 10 : 12 \quad \text{(authentic)}$$
$$H_0 = a : x_{\text{co-arith}} : x_{\text{arith}} : b = 6 : 7.2 : 9 : 12 \quad \text{(plagalic)}.$$

A_0 is the composition of A_3 and A_4. Musically realizing four-notes are quickly found: Let's take the F major chord in fifth position $c - f - a - c'$ for the authentic case and an E minor chord in tonal position $e - g - h - e'$.

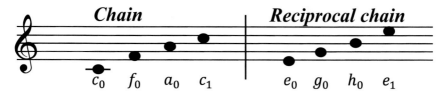

Very conveniently, we can read the property that one chord is reciprocal to the other. The step order has reversed.   ◄

For this diatonic octave canon D_{diat} another characteristic is revealed:

▶ **Completeness – Proposition**

The totality of all proportions which can be obtained from the diatonic octave canon $D_{\text{mus}}(1 : 2)$ by means of layering, fusion, and inversion of its proportions contained therein, is equal to that which can be obtained from the proportions of Senarius

$$S = 1 : 2 : 3 : 4 : 5 : 6$$

gains in the same way. And this interval set is the tone-, interval-, and scale structure known in music as **pure diatonic temperament or "just tuning"**. And exactly this **'completeness'** is meant by the attribute **'perfecta maxima'**.

The next two examples deal with the case of the fifth $2 : 3$ in the form $D_{\text{mus}}(6 : 9)$ and the octave fifth – the duodecimal $1 : 3$ – in the form $D_{\text{mus}}(6 : 18)$.

Note: At the same time, we want to demonstrate how these musical proportion-chains can be obtained in a simple way and without looking up the calculation formulas of the mean values in question: Only the simple arithmetic mean is to be given (all the rest results from the symmetry of our theorem) or even from the center properties of the geometric mean.

Let's start with the duodecimal $1 : 3 \cong 6 : 18$. In this proportion $6 : 18$ we have $x_{\text{arith}} = 12$. Now, according to the hyperbola principle of Theorem 2.5, the product is

$$x_{\text{arith}} * y_{\text{harm}} = 6 * 18 = 12 * y_{\text{harm}},$$

and therefore $y_{\text{harm}} = 9$. Now, we use that according to our theorem the subchain

$$y_{\text{harm}} : x_{\text{arith}} : y_{\text{co-harm}} = 9 : 12 : y_{\text{co-harm}}$$

is arithmetic. Therefore $y_{\text{co - harm}} = 15$. Now, finally, the hyperbolic property for $x_{\text{co - arith}}$ again provides the missing value: From the product

$$y_{\text{co-harm}} * x_{\text{co-arith}} = 6 * 18 = 108$$

then follows the value $x_{\text{co-arith}} = \frac{108}{15} = \frac{36}{5} = \frac{72}{10}$, and the entire proportion chain is determined without looking it up and only by means of the simplest mental arithmetic. We write this down in Example 3.10, where we still use an integer variant:

Example 3.10 The Duodecim Canon D_{mus} (6 : 18)

For the duodecime $1 : 3$ the musical chain of proportions is as follows

$$D_{mus}(6 : 18) \cong 6 : 7,2 : 9 : 12 : 15 : 18.$$

It has an integer-like variant

$$D_{\text{mus}}(10 : 30) \cong 10 : 12 : 15 : 20 : 25 : 30.$$

The duodecimal canon has a symmetry at least as perfect as the octave canon: In addition to the four arithmetic 2-step proportion chains $A_1, \ldots, A_4$ with their harmonic reciprocals $H_1, \ldots, H_4$, which are also partial chains of $D_{\text{mus}}(10 : 30)$. Specifically:

$$A_1 = 10 : 20 : 30 \ \text{with the Reciprocal} \ H_1 = 10 : 15 : 30,$$
$$A_2 = 15 : 20 : 25 \ \text{with the Reciprocal} \ H_2 = 12 : 15 : 20,$$
$$A_3 = 10 : 15 : 20 \ \text{with the Reciprocal} \ H_3 = 15 : 20 : 30,$$
$$A_4 = 20 : 25 : 30 \ \text{with the Reciprocal} \ H_4 = 10 : 12 : 15,$$

there is apparently an even 4-step arithmetic subchain

$$A_0 = 10 : 15 : 20 : 25 : 30,$$

whose 4-step reciprocal H_0 turns out to be a harmonic subchain of $D_{\text{mus}}(10 : 30)$,

$$H_0 = 10 : 12 : 15 : 20 : 30,$$

which perfectly confirms the reciprocity laws of Theorem 3.5. Musically, too, these 5-tone scales A_0 und H_0 can be verified by diatonically known models: A_0 is a spread diatonic major chord and indeed H_0 is a 5-tone minor chord:

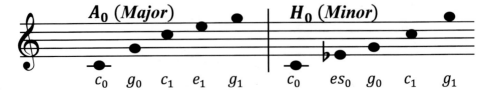

In this concrete tone example, we (as musicians) also see very nicely that both chains (chords) are reciprocal to each other: The interval sequence of the steps is exactly reversed. ◀

Finally, we consider the fifth canon: For the values of the musical proportion-chain of the fifth $D_{mus}(2:3)$, we use a more optimized integer $D_{mus}(20:30)$, and then, copying the preceding Example 3.10, we also find **by mental arithmetic** the values

$$x_{arith} = 25, y_{harm} = 24 \text{ and } y_{co\text{-}harm} = 26.$$

Then, according to the hyperbola principle or the principle of the centre of symmetry from Theorem 3.6, the value for the contra-arithmetic mean is given by

$$x_{co\text{-}arith} = \frac{20 * 30}{26} = \frac{300}{13} = 23\frac{1}{13}.$$

We have dispensed with a whole-numbered proportion chain in favor of smaller, more manageable numerical values and have thus obtained the proportion chain for the diatonic fifth canon:

Example 3.11 The Fifth Canon D_{mus} (2 : 3)

The musical chain of proportions of the fifth 2 : 3. Incorporated into the two outer proportions 20 : 30 and 6 : 9 is as follows:

$$D_{mus}(20:30) = 20 : 23\frac{1}{13} : 24 : 25 : 26 : 30$$

$$\cong 6 : 6\frac{12}{13} : 7,2 : 7,5 : 7,8 : 9 = D_{mus}(6:9).$$

Apart from its symmetry and the two arithmetic 2-step partial-chains and their harmonic reciprocals, however, it has no other chains of this kind.

However, this also shows that the number of at least two possible arithmetic and harmonic subchain pairs stated in the theorem cannot be increased already for the general case. ◀

These examples can now be continued in manifold ways and one can embark on journeys of discovery.

Question What symmetries (for example) does a musical diatonic canon-chain of the span 1 : 5 – i.e. of an overtone third – possess? A short hint: It is worthwhile to look at the proportion 1 : 5 in the form similar to this one

$$39 : 195 = (3 * 13) : (5 * 3 * 13)$$

to consider. For freaks, the task is then to show that the diatonic canon $n : 5n$ is an integer exactly, when n is a multiple of 39.

Finally, we would like to encourage you to practice this method for calculating diatonic canon chains, which is obtained by mental arithmetic. As nowhere else do the symmetrical relationships become more apparent than with this method for calculating all four musical modes, which is based on theory and accompanies it. We once again pin this method as a pioneering recipe:

Par-Coeur Method for Calculating the Diatonic Canon

$$D_{mus}(a,b) = a : x_{\text{co-arith}} \; : y_{\text{harm}} \; : x_{\text{arith}} \; : y_{\text{co-harm}} : b$$

For convenience, we'll assume the integer form $a = n$ and $b = m$, and then the procedure of determining the canon-chain by mental arithmetic (and without formula knowledge) works in the following steps:

1. First calculate the sum and product of the magnitudes, $n + m$ and nm.
2. Then calculate the arithmetic mean $x_{\text{arith}} = (n + m)/2$.
3. Then calculate the harmonic mean via $y_{\text{harm}} = nm/x_{\text{arith}}$.
4. Then calculate the contra-harmonic mean from this. To do this, we first determine the difference

$$d = x_{\text{arith}} - y_{\text{harm}}$$

and add these to the arithmetic mean, then $y_{\text{co - harm}} = x_{\text{arith}} + d$.
5. Now calculate the contra-arithmetic mean via $x_{\text{co - arith}} = nm/y_{\text{co - harm}}$.

So what was necessary to remember? Actually, we only have to calculate the arithmetic mean (the others are given to us by the Harmonia abstracta) and so we can calculate – par coeur, by heart and only by means of the multiplication tables of mental arithmetic and not entirely without pride – any diatonic canon chains.

Proportion Sequences of Babylonian Medieties

4

> *. . . let the heavens be kept in motion according to the sweetness of harmony – Aristotle (from [5], p. 127)*

In historical music theory, we constantly encountered the process of finding one or more additional magnitudes (i.e. tones) to two data – the magnitudes of an interval – in a way that the old data together with the new data formed proportion chains of a desired type of medieties. In Sect. 3.4 we have already described this process as a construct of various babylonian third proportions, as far as we could use this for the harmonia perfecta maxima.

Getting back to the basic example: This is how the proportion

Pythagorean (major) whole tone 8 : 9

turns into the arithmetic chain 8 : 9 : 10, by adding the value 10, which, as we have seen, expands the pythagorean system into a purely diatonic system. In addition to the two generating prime numbers 2 and 3 of the pythagorean system, the prime number 5 is added, from whom the major pure third 4 : 5 is born. If we complement this step with a reciprocal – and therefore harmonic – chain of proportions, we achieve the familiar musical chain of proportions that has been reduced to an integer form, the canon

$$30 : 36 : 40 : 45 : 50 : 60$$

of pure diatonics.

▶ The idea now is that this *process is restarted over and over again, while, instead of the start-data 8 and 9, any other ones of the gradually building up data-system are chosen.*

Mathematically, this is expressed in a way that along this chapter we will describe an analysis that embeds the most important historical medieties in a new mathematical theory for three differently conceptualized families of ad infinitum iterated mean-proportion chains $a : x : b$. Namely for

- the families of **iterated babylonian medieties** as nested iterations of the babylonian chains of proportions in the *interior* of the interval $[a, b]$,
- the families of the **third and higher proportions** as interpolation of progressing mean proportions into the exterior of the interval $[a, b]$,
- the families of '**contra-medieties**' as a bipartite series of iterated proportion chains belonging to the exponential division parameter series $(a/b)^n$.

In fact, a further multitude – better: a myriad – of medieties is achieved, which results from the fact that a third magnitude is always obtained from two existing data, perhaps even in a different way, so that these three data form the chain of medieties of a required architecture or a required type of medieties. If one again picks out two of these three data, then renewed averaging leads to further data. Hence, in some literatures, one encounters almost incalculable series of chains of medieties. Our approach arises from the urge to design an analytical concept for the most important of these iterations of various medietary proportions – occasionally called interpolations – in order to bring these things into a familiar modern form of our mathematics. First and foremost, we face the urge to gain a classification of all these processes, that shows the relations to the harmonia perfecta and its systematics.

In these iterations, it is above all the "inner" iteration of musical medieties into an infinite chain of musical proportions converging inwards towards the geometric mean as the centre, which holds an infinite copy of the theorems of Iamblichos and Nicomachus – and which therefore deserves only one name: the **harmonia perfecta infinita.**

But also, the contra-mediety series, in its composition as a bipartite infinite chain of proportions, harbors a captivating variety of inner symmetries of its babylonian musical sub-chain systems – likewise set forth in its harmonia perfecta infinita.

In our analysis, we, once again, make use of the uniform representation of all mean value proportions in the "piece-to-remainder"-form, and the mathematical analysis of these proportions then makes use of the analytical and structural properties of the "piece-to-rest"-proportion-function obtained in Theorem 3.2 of Sect. 3.1

$$y = f(x) = (x - a)/(b - x)$$

and its inverse, the mean value function

$$x = g(y) = a + \frac{y}{1+y}(b-a);$$

whose functional equation (see Theorem 3.1) again acts as a refined method describe the relationships of the mean symmetries with respect to the symmetries of the division parameters.

4.1   Higher Proportionals and Their Babylonian Mean Value Chains

As pointed out in the introduction to this chapter, we start from a given 1-level chain (proportion) $A = a_0 : a_1$, find a front or rear 3rd proportional of a given babylonian type, thus obtain a 2-level chain, select again a front or rear 1-level sub-chain, extend it again by a front or rear 3rd proportional, thus obtaining a 3-level chain and so on.

One can try to continue this process after each step – by choosing 1-step sub-chains in, even, extremely nested and manifold ways and thus reaches – in the older literature so called – **'higher proportional'**. These processes can be continued – perhaps – iteratively into infinity – knowing that this "secret science" is increasingly losing its practical significance within scale theory. But we are curious and will investigate the question of continuability:

Where does this end up?

We turn to the actual question, when a given proportion, respectively, a given babylonian chain of proportions can be extended iteratively – again and again – by front or back members in such a way, that either the babylonian type of the chain is preserved or another given type is obtained.

Agreement   For reasons of a technical description, we use for a generally given proportion chain $A = a_0 : a_1 : \ldots : a_n$ the two parameters of the 'initial' ratio increase or the 'initial' difference,

$$q = \frac{a_1}{a_0} \quad \text{respectively} \quad d = a_1 - a_0;$$

since they play a characteristic role in the case of our three chosen babylonian medieties.

Theorem 4.1 (Existence of Higher Babylonian Proportionals)
For all three types of babylonian medieties (geometric/arithmetic/harmonic), the
following characterizations, criteria and formulas about the existence of the iterations
of higher babylonian proportionals apply:

1. **Geometric chains:** A proportion chain $G = a_0 : a_1 : \ldots : a_n$ is geometric if and
   only if the recursion formulas

$$a_{k+1} = a_k * q \text{ for all } k = 0, \ldots, n-1$$

   apply. This recursive system can instantly be integrated into the system of direct
   calculations

$$a_k = a_0 * q^k \text{ for all } k = 0, \ldots, n-1$$

   whereas each member of the series can be specified immediately.
   Conclusion: Every proportion $a_0 : a_1$ can be unambiguously continued to a
   geometric chain

$$G_\infty = \ldots : a_{-m} : \ldots : a_{-1} : a_0 : a_1 : \ldots : a_n : \ldots$$

   progressing indefinitely on both sides. If the frequency measure of the start
   proportion is $q = a_1/a_0 > 1$, the following applies

$$a_{-m} \to 0 \text{ for } m \to \infty \text{ and } a_n \to \infty \text{ for } n \to \infty.$$

   For $q = a_1/a_0 < 1$ the convergence behavior is reversed – it is then preferred to
   consider the inverse proportion $a_1 : a_0$.
2. **Arithmetic chains**: A proportion chain $A = a_0 : a_1 : \ldots : a_n$ is arithmetic if the
   recursion formulas or equations apply:

$$a_{k+1} - a_k = d \quad \text{for all } k = 0, \ldots, n-1 \quad \text{(Recursion formulas)}$$
$$a_k = a_0 + kd \quad \text{for all } k = 0, \ldots, n-1 \quad \text{(System of equations)}$$

   Conclusion: Each proportion $A = a_0 : a_1$ (with $0 < a_0 < a_1$) can be continued on
   the left-hand side at most finitely (m) times – on the right-hand side, however, as
   often as desired – to an altogether infinitely progressing arithmetic chain

(continued)

Theorem 4.1 (continued)
$$A_\infty = a_{-m} : \ldots : a_{-1} : a_0 : a_1 : \ldots : a_n : \ldots$$

The left-hand continuation terminates after a finite number of steps if for a negative index k the expression $a_0 + kd \leq 0$ arises.

3. **Harmonic chains:** A proportion chain $H = a_0 : a_1 : \ldots : a_n$ is harmonic if and only if the recursive (and mutually equivalent) equations

$$a_{k+2} = \frac{a_k a_{k+1}}{2a_k - a_{k+1}} \quad \text{for all} \ k = 0, \ldots, n-2$$

$$\Leftrightarrow a_k = \frac{a_{k+1} a_{k+2}}{2a_{k+2} - a_{k+1}} \quad \text{for all} \ k = 0, \ldots, n-2$$

are fulfilled. The second equation is simply the rearrangement of the first equation to the variable a_k.

Conclusion: Every proportion $A = a_0 : a_1$ (with $0 < a_0 < a_1$) can be continued on the right-hand side at most finitely often (*m times*) – to the left, however, as often as desired to a harmonic chain

$$H_\infty = \ldots : a_{-n} : \ldots : a_{-1} : a_0 : a_1 : \ldots : a_m$$

Proof: To (1): Right away from the definition of a geometrical chain of proportions follows that the quotient of successive magnitudes is always the same – concludes the recursion formula. The equivalent resolution of the recursion to an equation is obtained by recognizing in turn:

$$a_1 = a_0 q \Rightarrow a_2 = a_1 q = a_0 q^2 \Rightarrow a_3 = a_2 q = a_0 q^3 \Rightarrow \ldots$$

(the proof is then completed by the mathematically correct application of the principle of 'complete induction'). Conversely, if the magnitudes a_k for all integer indices k according to this equation are

$$a_k = a_0 q^k$$

then the bipartite unrestricted chain G_∞ of the ordered magnitudes is geometric, since the recursive formula always applies. Three successive magnitudes each form a 2-step geometric chain, as required.

To (2): The relation between recursion and equation is trivial; one adds up the difference d from magnitude to magnitude – thus the indicated balance arises. To the left, the process breaks off when the sense barrier of the positivity of the magnitudes is breached, when, although still

$$a_0 - md > 0, \text{ but then the inequality } a_0 - (m+1)d \leq 0$$

is reached, which will occur at some point if the difference parameter d is positive.

To (3): If we did not have a 'theory' at hand, the iterative description together with its decision criteria could demand quite unpleasant calculations: Thus, however, we enjoy the relation **that the reciprocal of a harmonic chain is arithmetic,** and we can therefore use the trivial statement (2). The actual recursion formulas, however, are the direct calculation of a_{k+2} as the posterior and a_k as the anterior harmonic 3rd proportional of the sub-chain $a_k : a_{k+1} : a_{k+2}$, where, after all, a_{k+1} is the harmonic mean of its neighbors a_k and a_{k+2}. The solutions of the mean equation

$$a_{k+1} = 2a_k a_{k+2}/(a_k + a_{k+2})$$

to a_{k+2} and to a_k are then exactly the recursion formulas given in the theorem; we had already calculated the second of these in Theorem 3.4. The first would result from an exchange of the indices anyway, since the formula of the harmonic mean is symmetrical with respect to the two external magnitudes a_k and a_{k+2}. The continuability of the harmonic chain results from that of the arithmetic reciprocals; and because in this reversal the order of the proportions is reversed, the described statements are consequences of the arithmetic case 2).

4.2   Sequences of Contra-Medieties

We will now describe a generalization – extensively discussed in the nineteenth and twentieth centuries – of harmonic and arithmetic chains of proportions and their mean values. The examples of the classic medieties and their piece-to-rest-proportions already show that many of them have the formal structure

$$(x-a):(b-x) \cong b^n : a^n \cong q^n : 1$$

And then the exponent (n), along with the frequency measure q, is the characteristic parameter of the mediety in question. Thus $n = 0$ belongs to the arithmetic mean, $n = -1$ to the harmonic mean – and if we would also allow fractional exponents, $n = -1/2$ would be the appropriate parameter to the geometric mean.

Definition 4.1 (The Sequence of Contra-Medieties)

Let $a < b$, then for $n = 0, \pm 1, \pm 2, n \in \mathbb{Z}$, we define the family of proportion chains $a : x : b$ by the division rule (proportion)

$$(x - a) : (b - x) \cong b^n : a^n \quad - \text{ Contra} - \text{mediety proportions.}$$

By setting again $q = b/a$, in modern language this family of proportions is parametrized over the parameter q and the exponent n, so that we have a family of proportions to an **exponential** division parameter sequence q^n

$$(x - a)/(b - x) = q^n \text{ with } n = 0, \pm 1, \pm 2, \ldots$$

With x_n we denote the – according to Theorem 3.2 – uniquely existing solution of this proportion equation to the exponent n; therefore applies

$$(x_n - a)/(b - x_n) = q^n$$

with $a < x_n < b$. In such manner we obtain the bipartite **contra-medieties sequence**

$$\ldots x_{-3}, \, x_{-2}, \, x_{-1}, \, x_0, \, x_1, \, x_2, \, x_3, \, \ldots,$$

whose magnitudes are all distributed in the open real interval$]\,a, b\,[$, whilst being obedient to the monotonicity of the two functions

$$y = f(x) = (x - a)/(b - x) \text{ and } x = g(y) = \frac{y}{1 + y}(b - a),$$

see Theorem 3.1. In addition, for – in principle arbitrary – real exponents n, an associated unique mediety x_n can be found, which is the solution of the equation

$$(x_n - a)/(b - x_n) = q^n.$$

So for $n = -1/2$, the mediety would be $x_{-1/2} = z_{\text{geom}}(a, b)$.

Since the sequence q^n is an **exponential sequence** regarding the discrete variable n, we could also say '**exponential proportion family**'. However, the basic mathematical Theorem 3.1, which describes the symmetry between the medieties and their division parameters q^n, suggests us to prefer the positional relation, which leads to the contra terms, in the notation. Finally, in Sect. 3.5 we studied in more detail the symmetric positioning to the geometric mean as a function of the exponential parameter n.

We already know – as mentioned – some representatives of this family:

Example 4.1 Musical Medieties in the Contra-Medieties Sequences

The four mean values of the diatonic canon (arithmetic, harmonic, contra-arithmetic and contra-harmonic), which are considered as "musical medieties", are members of the contra-mediety sequences:

n	q^n	Mean value formula	Alternative formula	Proportion chain
$n = 0$	1	$x_0 = \frac{a+b}{a^0+b^0}$	$x_0 = \frac{1}{2}(a+b)$	Arithmetic
$n = -1$	a/b	$x_{-1} = \frac{a^0+b^0}{a^{-1}+b^{-1}}$	$x_{-1} = \frac{2ab}{a+b}$	Harmonic
$n = 1$	b/a	$x_1 = \frac{a^2+b^2}{a+b}$	$x_1 = \frac{a^2+b^2}{a+b}$	Contra-harmonic
$n = -2$	a^2/b^2	$x_{-2} = \frac{a^{-1}+b^{-1}}{a^{-2}+b^{-2}}$	$x_{-2} = ab\frac{a+b}{a^2+b^2}$	Contra-arithmetic

---

The following theorem describes not only the formulas for the calculation of the mean values x_n, which can be solved according to the simple equation of determination

$$f(x_n) = \frac{(x_n - a)}{(b - x_n)} = q^n \Leftrightarrow x_n = g(q^n) = a + \frac{q^n}{1 + q^n}(b - a)$$

but it also gives information about their position in relation to each other, their distribution in the real interval $[a, b]$ and about some remarkable symmetries among each other – with the result that there is a controlling order in this hodgepodge of mean values, which manifests itself in the fact that we find three ordered families with an infinite number of geometric, an infinite number of arithmetic and an infinite number of harmonic subproportion chains. Thereby are

all geometric chains centered around the geometric mean $z_{geom} = \sqrt{ab}$,
all arithmetic chains centered around the arithmetic mean $x_{arith} = (a + b)/2$,
all harmonics chains centered around the harmonic mean $y_{harm} = 2ab/(a + b)$.

This order also underpins our notation of the series as a contra-medieties sequences – for this is what it then is in three respects. This order also represents a striking connection of analysis with historical mediety proportions and their musical interpretations.

Theorem 4.2 (Analysis for the Sequence of Contra-Medieties)

The exponential proportion family of contra-medieties

(continued)

Theorem 4.2 (continued)

$$(x-a) : (b-x) \cong b^n : a^n$$

defines a bipartite sequence of mean values

$$\ldots x_{-3}, x_{-2}, x_{-1}, x_0, x_1, x_2, x_3, \cdots = (x_n)_{n \in \mathbb{Z}},$$

for which we can give the following mean value formulas, position relations and symmetries:

1. **Mean value formulas:** Let $n = 0 \pm 1, \pm 2, \ldots$ be any positive or negative proportion parameter, and let $m = -n$ be set for notational convenience. The following formulas are equivalent:

   (a) $(x_n - a)/(b - x_n) = q^n = b^n/a^n$,

   (b) $x_n = a \frac{1+q^{n+1}}{1+q^n} = b \frac{1+q^{m-1}}{1+q^m}$.

   If we express the **mean value formulas** in (b), which have the frequency factor q, again by the data a and b, we get the two interesting variants:

   $$x_n = \frac{a^{n+1} + b^{n+1}}{a^n + b^n} \quad \text{and} \quad x_n = ab \frac{a^{m-1} + b^{m-1}}{a^m + b^m}.$$

   The former of these is called the '**arithmetic**' form and the latter the '**harmonic**' form. The harmonic form is especially preferred when $n < 0$ is and thus inevitably $n = -m$ is written with a positive m.

2. **Positional relationships:** First applies the order

   $$n < m \iff a < x_n < x_m < b \quad \text{(for all integers } n, m\text{)}.$$

   Moreover, we have the monotonous convergence

   $$x_n \nearrow b \text{ (for } n \to +\infty\text{) and } x_n \searrow a \text{ (for } n \to -\infty\text{)}.$$

3. **Symmetry formulas:** For all exponents $n \in \mathbb{Z}$ the following equations are valid

   (a) $x_n * x_{-(n+1)} = ab = z_{\text{geom}}^2 \iff x_{-1/2} = z_{\text{geom}}(a, b) = z_{\text{geom}}(x_{-(n+1)}, x_n)$

   (b) $x_n - x_0 = x_0 - x_{-n} \iff x_0 = x_{\text{arith}}(a, b) = x_{\text{arith}}(x_{-n}, x_n)$

(continued)

Theorem 4.2 (continued)

(c) $(x_{-1} - x_{1-n})/(x_{-1+n} - x_{-1}) = x_{-1-n}/x_{-1+n}$

$\Leftrightarrow x_{-1} = y_{harm}(a,b) = y_{harm}(x_{-1-n}, x_{-1+n})$

4. **Proportions of the mean values:** From the symmetry formulas, the following ascending, 2-step, babylonian proportion chains emerge:

For all positive integers n the following proportion chains have the properties:

(a) $G_n = x_{-(n+1)} : x_{-(1/2)} : x_n$ are geometric,

(b) $A_n = x_{-n} : x_0 : x_n$ are arithmetic,

(c) $H_n = x_{-1-n} : x_{-1} : x_{-1+n}$ are harmonic.

Special, the classical babylonian mean-value – magnitudes

$$x_{-(1/2)} = z_{geom} = z_{geom}(a,b) = \sqrt{ab} \equiv \text{the geometric mean,}$$

$$x_0 = x_{arith} = x_{arith}(a,b) = \frac{1}{2}(a+b) \equiv \text{the arithmetic mean,}$$

$$x_{-1} = y_{harm} = y_{harm}(a,b) = \frac{2ab}{a+b} \equiv \text{the harmonic mean,}$$

to the fixed data a and b are apparently **universal symmetry centers of** all corresponding sub-chains G_n, A_n and H_n.

The scheme of Fig. 4.1 shows the babylonian symmetry centers of the contra-medieties sequences

Comment The harmonic form of the mean value formulas is used in a way that no negative powers occur in the fractions for negative parameters n. Otherwise double fractions would occur in these cases. An example:

For $n = -2$ and thus for $m = 2$ we read off the two plots for the 'contra-arithmetic' mean:

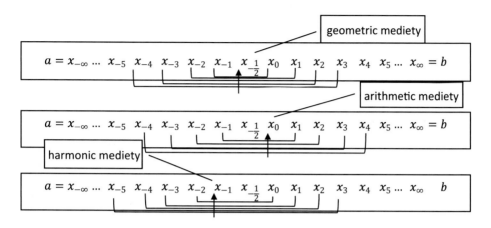

Fig. 4.1 Scheme of contra-medieties sequences

$$x_{-2} = \frac{a^{-1} + b^{-1}}{a^{-2} + b^{-2}} = \frac{\dfrac{1}{a} + \dfrac{1}{b}}{\dfrac{1}{a^2} + \dfrac{1}{b^2}} \quad (\equiv \text{ arithmetic formula}),$$

$$x_{-2} = ab\frac{a^1 + b^1}{a^2 + b^2} = ab\frac{a + b}{a^2 + b^2} \quad (\equiv \text{ harmonic formula}).$$

Proof We obtain the first statement (1) directly from the mean value formula of Theorem 3.1. Here we derive the first of the two (frequency factor) forms of the mean value formula, namely

$$x_n = a\frac{1 + q^{n+1}}{1 + q^n}.$$

The 'alternative formula' is gained simply by substituting $m = -n$ and expanding the fraction by the factor. Consequently we obtain the equation sequence

$$a\frac{1 + q^{n+1}}{1 + q^n} = a\frac{1 + q^{1-m}}{1 + q^{-m}} = a\frac{1 + q^{1-m}}{1 + q^{-m}} * \frac{q^m}{q^m} = a\frac{q^m + q}{q^m + 1} = aq\frac{q^{m-1} + 1}{q^m + 1} = b\frac{q^{m-1} + 1}{q^m + 1}.$$

Both mean value formulas in the a-, b-data result by inserting $q = b/a$ in the respective frequency factor forms with subsequent expansion of the resulting fractions with a^n, respectively a^m.

From the arithmetic form, the harmonic form also follows in a direct way: If we set $n = -m$, we get immediately, and after extension of the fraction with $(ab)^m$

$$x_n = \frac{a^{n+1} + b^{n+1}}{a^n + b^n} = \frac{a^{1-m} + b^{1-m}}{a^{-m} + b^{-m}} = \frac{a^{1-m} + b^{1-m}}{a^{-m} + b^{-m}} \frac{(ab)^m}{(ab)^m}$$

$$= \frac{ab^m + ba^m}{b^m + a^m} = \frac{abb^{m-1} + aba^{m-1}}{b^m + a^m} = ab \frac{b^{m-1} + a^{m-1}}{b^m + a^m} = ab \frac{a^{m-1} + b^{m-1}}{a^m + b^m}.$$

Regarding statement (2), we observe that for the positive exponential sequence q^n, the well-known monotonicity and limit considerations apply: Since $q > 1$ holds, we get the patterns

$$q^n < q^m \text{ for } n < m$$
$$q^n \nearrow \infty \text{ (for } n \to +\infty) \text{ and } q^n \searrow 0 \text{ (for } n \to -\infty).$$

(The arrows $\nearrow$ and $\searrow$ express a strictly monotonically increasing and a strictly monotonically decreasing striving behaviour). Statement (2) follows directly from the mean value monotonicity of Theorem 3.1.

We verify statements (3) by mere recalculation:

To (a): It is first of all interesting that the relation

$$x_n * x_{-(n+1)} = ab$$

must actually follow exactly from the functional equation of the proportion function of Theorem 3.1: Namely, x_n and $x_{-(n+1)}$ are solutions of the equation

$$f(x_n) = q^n \text{ and } f(x_{-(n+1)}) = q^{-(n+1)} = 1/(q * q^n).$$

The division parameters therefore satisfy the condition

$$f(x_n) * f(x_{-(n+1)}) = 1/q = a/b,$$

So that we can conclude according to the functional equation from Theorem 3.1 – but also with the 'magic formula' of Theorem 3.7 – that

$$x_n * x_{-(n+1)} = ab$$

must apply. However, for control (whether all our formulas are correct), we can also 'recalculate' the alleged relation:

$$\begin{aligned}
x_n * x_{-(n+1)} &= a\frac{1+q^{n+1}}{1+q^n} * a\frac{1+q^{-n}}{1+q^{-(n+1)}} = a^2\frac{1+q^{n+1}}{1+q^n} * \frac{1+q^{-n}}{1+q^{-(n+1)}} * \frac{q^n}{q^n} \\
&= a^2\frac{1+q^{n+1}}{1+q^n} * \frac{q^n+1}{q^n+q^{-1}} = a^2\frac{1+q^{n+1}}{(1+q^{n+1})q^{-1}} = a^2 q = ab.
\end{aligned}$$

This is formula (a) – where we note that if we use both versions of the formula (arithmetic and harmonic), the assertion can even be read off – without calculation. We consider formula (b) like this:

$$\begin{aligned}
x_0 - x_{-n} &= a\frac{1+q}{1+q^0} - a\frac{1+q^{-n+1}}{1+q^{-n}} = a\frac{1+q}{2} - a\frac{1+q^{-n+1}}{1+q^{-n}} * \frac{q^n}{q^n} \\
&= a\frac{1+q}{2} - a\frac{1+q^{-n+1}}{1+q^{-n}} * \frac{q^n}{q^n} = a\frac{1+q}{2} - a\frac{q^n+q}{q^n+1} = a\frac{1-q^n-q+q^{n+1}}{2(1+q^n)}, \\
x_n - x_0 &= a\frac{1+q^{n+1}}{1+q^n} - a\frac{1+q}{2} = a\frac{2+2q^{n+1}-(1+q)(1+q^n)}{2(1+q^n)} = a\frac{1+q^{n+1}-q-q^n}{2(1+q^n)},
\end{aligned}$$

and both expressions correspond to each other.

Although 'elementary', the third formula (c) unfortunately requires a more complex calculation: Cross multiplication yields the equation equivalent to the formula

$$(x_{-1} - x_{-1-n})/x_{-1-n} = (x_{-1+n} - x_{-1})/x_{-1+n},$$

which, by shortening and summarizing, turns into the equation to be proved

$$x_{-1}\left(\frac{1}{x_{-1-n}} + \frac{1}{x_{-1+n}}\right) = 2.$$

By putting the mean formulas in the frequency factor form and noting that the factor a truncates out anyway, this is what remains:

$$\frac{1+1}{1+q^{-1}}\left(\frac{1+q^{-1-n}}{1+q^{-n}}+\frac{1+q^{-1+n}}{1+q^n}\right)=2$$

$$\Leftrightarrow \frac{2q}{1+q}\left(\frac{1+q^{-1-n}}{1+q^{-n}}*\frac{q^{n+1}}{q^n q}+\frac{1+q^{-1+n}}{1+q^n}\right)=2$$

$$\Leftrightarrow \frac{2q}{1+q}\left(\frac{1+q^{1+n}}{1+q^n}*\frac{1}{q}+\frac{1+q^{-1+n}}{1+q^n}\right)=2$$

$$\Leftrightarrow \frac{2q}{q(1+q)(1+q^n)}(1+q^{1+n}+q(1+q^{-1+n}))=2$$

$$\Leftrightarrow \frac{2}{(1+q)(1+q^n)}(1+q^{1+n}+q+q^n)=\frac{2}{(1+q)(1+q^n)}((1+q)+q^n(1+q))=2.$$

We found the last step by target-driven factoring, so that the whole denominator truncates away and assertion (3) is proved.

The proportion chain of the contra-medieties and their own harmonia perfecta will be discussed in Sect. 4.5.

4.3   Unilateral Iterations of Babylonian Medieties

We will describe a process, which consists of an iterative construction of permanent new extractions of Babylonian mean values x_{arith}, y_{harm}, z_{geom} and that is 'inwardly' nested.

As so often, we start from two initial data $a < b$, which preferably stand for a musical interval $[a, b]$ with the frequency factor $q = b/a$, We mention again that the babylonian mean values for given data a and b in the known arrangement

$$a < y_{\text{harm}}(a,b) \left(< z_{\text{geom}}(a,b) \right) < x_{\text{arith}}(a,b) < b$$

are ordered – and – most importantly – that also the inner chain of proportions

$$y_{\text{harm}}(a,b) : z_{\text{geom}}(a,b) : x_{\text{arith}}(a,b)$$

itself is geometric again – a result of the harmonia perfecta maxima babylonica, respectively of the Iamblichos/Nicomachus theorem. This means that $z_{\text{geom}}(a,b)$ is also the geometric mean of $a_1 := y_{\text{harm}}(a,b)$ and $b_1 := x_{\text{arith}}(a,b)$. Consequently, with $a_0 = a$, $b_0 = b$ the magnitude series

$$a_0 < a_1 < z_{\text{geom}}(a,b) < b_1 < b_0$$

is ordered this way and $z_{\text{geom}}(a,b)$ is the center of symmetry of the 3-step proportion chain

$$a_0 : a_1 : b_1 : b_0,$$

which therefore – by the principle of symmetry of the hyperbola of Archytas according to Theorem 3.6 – is then also a geometrical chain of proportions. If we take – as a second step, respectively as a second iteration – these two values a_1 and b_1 as **new initial data,** to whom we again form the babylonian medieties 'harmonic' (a_2) and 'arithmetic' (b_2), we obtain altogether the medieties arrangement

$$(a=:)\,a_0 < a_1 < a_2 < z_{\text{geom}}(a,b) < b_2 < b_1 < b_0\,(:=b),$$

because the geometric mean z_{geom} has remained unchanged, and is now simultaneously the geometric mean of all three data pairs (a_0, b_0), (a_1, b_1) and (a_2, b_2). Now, as the next iteration, we select a_2, b_2 as the new initial data, obtain its harmonic and arithmetic mean a_3, b_3 and once again z_{geom} is the common geometric mean of a_2, b_2 and a_3, b_3. The resulting series is then nested in the following way

$$a_0 < a_1 < a_2 < a_3 < z_{\text{geom}} < b_3 < b_2 < b_1 < b_0$$

It is immediately clear how this iterative process continues. One should note that due to the fact that all inequalities are strict, all these new data are pairwise different from each other, so that in this way we get two strictly monotone sequences of medieties

$$a = a_0 < a_1 < a_2 < a_3 < \cdots < z_{\text{geom}} \quad \text{(sequence of iterated harmonic means)},$$
$$z_{\text{geom}} < \cdots < b_3 < b_2 < b_1 < b_0 = b \quad \text{(sequence of iterated arithmetic means)}$$

From these outcomes we can right away conclude:

1. The sequence $(a_n)_{n \in \mathbb{N}}$ of iterated harmonic medieties is monotonically increasing and bounded upwards by the geometric mean z_{geom} – thus also convergent according to the completeness theorem of the real numbers.
2. The sequence of iterated arithmetic medians $(b_n)_{n \in \mathbb{N}}$ is monotonically decreasing and this time bounded downward by the geometric mean z_{geom} – thus also convergent.
3. According to the symmetry principles from Theorem 3.6 – but also according to the theorem of Iamblichos/Nicomachus – the geometric mean z_{geom} of a and b is simultaneously the geometric mean **for all iterated data pairs** (a_n, b_n). So for all $n \in \mathbb{N}$ the mean equations

$$a_n * b_n = (z_{\text{geom}})^2 = ab,$$

are valid **and all data pairs (a_n, b_n) lie on the hyperbola of Archytas $y = ab/x$.** Therefore the simple relations

$$a_n = ab/b_n \text{ respectively } b_n = ab/a_n,$$

exist between both sequences, so that they are basically reciprocal to each other. If one knows certain properties of one sequence, one can immediately conclude corresponding statements for the other.

We now firstly develop a small formula world of these iterations and summarize these considerations with a number of others in a theorem:

Theorem 4.3 (Analysis of the Inner Babylonian Medieties Sequences)

1. **The recursive system of the sequences of the arithmetic respective harmonic means**

    Given an interval $a < b$, the sequence of iterated babylonian medieties consists of the two sub-sequences

    $(a_n)_{n \in \mathbb{N}}$ ($\equiv$ iterated harmonic means), $(b_n)_{n \in \mathbb{N}}$ ($\equiv$ iterated arithmetic means),

    which, according to our preceding construction, satisfy the following recursive system of equations: With the initial value anchoring

    $$a_0 = a, \ b_0 = b$$

    the sequences $(a_n)_{n \in \mathbb{N}}$ and $(b_n)_{n \in \mathbb{N}}$ are recursively calculated in the following several ways:

(A) **Coupled recursion system:** A first relation provides the system

$$a_{n+1} = \frac{2a_n b_n}{a_n + b_n} \text{ and } b_{n+1} = \frac{1}{2}(a_n + b_n), \ (\text{for } n = 0, 1, 2 \ldots),$$

whereas a_{n+1} is the harmonic, respectively b_{n+1} the arithmetic mean of the previous data a_n, b_n. These equations are a system of two coupled recursion formulas. Because the geometric mean z_{geom} of a and b *is* also the geometric mean of all data pairs (a_n, b_n), the **mean equation**

$$a_n * b_n = (z_{\text{geom}})^2 = ab$$

holds for all n. All pairs of data (a_n, b_n) lie on the hyperbola of Archytas.

By these mean value equations the system (A) simplifies and gets the form

(continued)

Theorem 4.3 (continued)

(B) Simplified recursion system:

$$a_{n+1} = \frac{2ab}{a_n + b_n} \text{ and } b_{n+1} = \frac{1}{2}(a_n + b_n) \text{ with the initial condition } a_0 = a, \ b_0 = b$$

and can even be completely decoupled to the new system because of $a_n = ab/b_n$

(C) Decoupled recursion system:

$$a_{n+1} = \frac{2ab}{a_n^2 + ab}a_n \text{ and } b_{n+1} = \frac{1}{2}\left(b_n + \frac{ab}{b_n}\right),$$

And the decoupled system (C) is equivalent to the coupled initial system (A).

2. **The convergence of the arithmetic-harmonic inner medieties**

The two sequences (a_n, b_n) of the iterated arithmetic-harmonic medieties are monotone, bounded – hence convergent – sequences,

$$a = a_0 < a_1 < a_2 < a_3 < \cdots < z_{\text{geom}} < \cdots < b_3 < b_2 < b_1 < b_0 = b.$$

The iterated harmonic medieties sequence (a_n) is strictly monotonically increasing and bounded upwards. The iterated arithmetic medieties sequence (b_n) is strictly monotonically decreasing and bounded downwards. The expected behavior of the monotonic convergence to the geometric mean now applies to the limit values z_{geom},

$$a_n \nearrow z_{\text{geom}} \text{ and } b_n \searrow z_{\text{geom}} \ (\text{for } n \to \infty).$$

Musical interpretation: The iterated intervals $[a_n, b_n]$ become smaller and smaller as the series index n increases, so that – in musical terms – the frequency factor $b_n/a_n \to 1$ strives, which in turn means,

'that the tonal intervals $[a_n, b_n]$ converge towards the unison (1 : 1)'.

Proof to (1): The recursion system (A) initially corresponds exactly to the intended constructive procedure of iterated medietary extraction. By then applying Nicomachus' theorem recursively ('step by step'), we obtain the all-simplifying relation

$$a_n * b_n = ab$$

for all indices n, from which the simplified form (B) of the coupled equations results. One can now insert the explicit forms $a_n = ab/b_n$, respectively $b_n = ab/a_n$, alternately, and thus get the totally decoupled recursions (C). As for the monotone position and the total arrangement, this also follows from Theorem 3.3 about the arrangement of babylonian medieties on given data together with the at all times important observation that the geometric mean for all pairs of data a_n, b_n is always constantly the mediety $z_{\text{geom.}}$

Proof of (2): We now present a very nice and elegant proof of the assertion that both monotone sequences of medieties also have the geometric mean as a common limit – and not only as a bound.

Let us consider the sequence $(b_n)_{n \in \mathbb{N}}$ of iterated arithmetic medieties. Because of the monotonicity and the boundedness of this series it follows that it converges. This is one of the fundamental statements of the whole analysis, equivalent to the completeness of the set of real numbers $\mathbb{R}$. Thus it holds that there is a number b_∞ such that

$$b_n \searrow b_\infty \text{ for } n \to \infty \quad - \text{ that is } b_\infty = \lim_{n \to \infty} b_n$$

applies. Now why is $b_\infty = z_{\text{geom}}$? Well, we can see that due to the following trick: We obviously have the following chain of inequalities:

$$z_{\text{geom}} < b_{n+1} = \frac{1}{2}(a_n + b_n) < \frac{1}{2}(z_{\text{geom}} + b_n) \iff 2z_{\text{geom}} < 2b_{n+1} < z_{\text{geom}} + b_n.$$

If this inequality holds for all series members, meaning also for the limit, where at most the case of equality can occur. Thus we have the weak inequality

$$2z_{\text{geom}} \leq 2b_\infty \leq z_{\text{geom}} + b_\infty.$$

From the first inequality follows immediately the inequality

$$z_{\text{geom}} \leq b_\infty,$$

and from the second we have the other inequality

$$b_\infty \leq z_{\text{geom}},$$

from whom altogether the equality

$$b_\infty = z_{\text{geom}}$$

follows as desired. For the likewise existing limit a_∞ of the monotonically growing series $(a_n)_{n \in \mathbb{N}}$ of the iterated harmonic medieties, which is bounded upwards by the geometric mean z_{geom}, then holds due to the relation

$$a_n * b_n = ab$$

also for their limit values a_∞ and b_∞ the equation

$$a_\infty * b_\infty = ab.$$

And now we put in the value for b_∞, consider the square-rectangle-formula for z_{geom}, and immediately the desired result follows

$$a_\infty = ab/b_\infty = z_{\text{geom}}{}^2 / z_{\text{geom}} = z_{\text{geom}}.$$

Hence, both sequences converge to the universal geometric mediety $z_{\text{geom}}(a, b)$.

Effort and Consolation *All in all, we have already presented and completed a respectable mathematical workload; not everything runs only on the level of the 'ratio calculation', the fraction calculus. However, in the last* Sect. 4.5 *of this chapter, we will reap the fruits of these arithmetical efforts – but also of their analytical mathematics – by presenting the double-infinite proportion chain*

$$B_{\text{mus}} = a_0 : a_1 : a_2 : \ldots : a_m : \ldots : \left(z_{\text{geom}} \right) \ldots : b_n : \ldots : b_2 : b_1 : b_0$$

in terms of their inner symmetries, encountering the harmonia perfecta maxima at every turn.

Remark: The Normalized Logistic Recursion System

By the way, there are also very interesting relations of the recursion system (A) to logistics: The recursion system can be brought to a concise normal form for this purpose: With the universal constant factor $1/z_{\text{geom}} = 1/\sqrt{ab}$ one defines the new quantities

$$y_n := \frac{1}{z_{\text{geom}}} a_n \quad \text{and} \quad x_n := \frac{1}{z_{\text{geom}}} b_n.$$

This 'normalization' corresponds to the standard situation that $a * b = 1$ applies. For then the coupled system (A) for a_n and b_n (where $n = 0, 1, 2, \ldots$ always holds) is equivalent to the new system

(D) Normalized recursion system

$$y_{n+1} = \frac{2y_n x_n}{y_n + \frac{1}{y_n}} \quad \text{and} \quad x_{n+1} = \frac{1}{2}\left(x_n + \frac{1}{x_n}\right),$$

and the simplified recursion system (B) for a_n and b_n is equivalent to the system

(E) Logistic recursion system

$$y_{n+1} = \frac{2}{y_n + \frac{1}{y_n}} \quad \text{and} \quad x_{n+1} = \frac{1}{2}\left(x_n + \frac{1}{x_n}\right)$$

with the anchoring $y_0 = \sqrt{a/b}$ and $x_0 = \sqrt{b/a}$, i.e. $x_0 * y_0 = 1$, and then this logistic system is again equivalent to the system

(F) Normalized logistic recursion system

$$x_{n+1} = \frac{1}{2}\left(x_n + \frac{1}{x_n}\right) \quad \text{with} \quad x_0 = \sqrt{b/a} \quad \text{and} \quad y_n * x_n = 1 \text{ for all } n \geq 0.$$

Furthermore, it is now valid altogether that this normalized logistic system (F) is equivalent to the coupled initial system (A). We show this in a short excursion into the world of arithmetics: First, the equivalence of both formula systems

$$\left(a_{n+1} = \frac{2a_n b_n}{a_n + b_n}, \ b_{n+1} = \frac{1}{2}(a_n + b_n)\right) \text{ and } \left(y_{n+1} = \frac{2y_n x_n}{y_n + (1/y_n)}, \ x_{n+1} = \frac{1}{2}\left(x_n + \frac{1}{x_n}\right)\right)$$

is clear. Less clear is the equivalence of the normal form (F) with the above system formulas (A) because the recursion formula for the sequence y_n is omitted in favor of the condition that $y_n * x_n = 1$ with the starting condition $x_0 = \sqrt{b/a}$. So it has to be proved more exactly, that the both systems

$$\left(y_{n+1} = \frac{2}{y_n + (1/y_n)}, \ x_{n+1} = \frac{1}{2}\left(x_n + \frac{1}{x_n} \right) \text{ with } y_0 = \sqrt{a/b} \text{ and } x_0 = \sqrt{b/a} \right)$$

$$\left(y_n * x_n = 1, \ x_{n+1} = \frac{1}{2}\left(x_n + \frac{1}{x_n} \right) \text{ with } x_0 = \sqrt{b/a} \right)$$

are equivalent. This means, however, that if $x_{n+1} = \frac{1}{2}\left(x_n + \frac{1}{x_n} \right)$ was fulfilled with $x_0 = \sqrt{b/a}$, then also the equivalence

$$\left(y_{n+1} = \frac{2}{y_n + (1/y_n)} \text{ with } y_0 = \sqrt{a/b} \right) \Leftrightarrow (y_n * x_n = 1)$$

would exist. This proof is in the following carried out with complete induction:

"$\Rightarrow$": According to assumption follows $y_0 * x_0 = 1$. We now assume that for a natural parameter n the equation $y_n * x_n = 1$ is correct and show that this is also true for the subsequent parameter $n + 1$ – this is, after all, the essence of the ‚induction principle'. It is then according to this induction assumption $y_n = 1/x_n$ and thus also because of $x_n = 1/y_n$ easy to see that

$$y_{n+1} * x_{n+1} = \frac{2}{y_n + \left(\frac{1}{y_n} \right)} * \frac{1}{2}\left(x_n + \frac{1}{x_n} \right) = \frac{x_n + y_n}{x_n + y_n} = 1$$

holds. Since the condition $y_0 * x_0 = 1$ is valid, according to this recursive argument it is firstly valid for $n = 1$, then for $n = 2$, and so on.

"$\Leftarrow$": Because $y_n * x_n = 1$ for all parameters $n \geq 0$, the equation $y_{n+1} = 1/x_{n+1}$ holds as well as simultaneously the equation $x_n = 1/y_n$. Then by inserting the recursion formula for the sequence x_n follows the equation

$$y_{n+1} = \frac{1}{\frac{1}{2}\left(x_n + \frac{1}{x_n} \right)} = \frac{2}{x_n + \frac{1}{x_n}} = \frac{2}{\frac{1}{y_n} + y_n},$$

as desired. Therefore we can replace the counter 2 with $2a_n b_n$ in the simplified form of the system for a_n and b_n, and thus achieve complete equivalence.

Finally, we remark on the system equivalence that, bypassing the simplifying normalization of both sequences by virtue of the uniform multiplication by the reciprocal geometric mean, we can also state:

The coupled recursion system

$$a_{n+1} = \frac{2a_n b_n}{a_n + b_n} \quad \text{and} \quad b_{n+1} = \frac{1}{2}(a_n + b_n) \quad \text{with the initial condition} \quad a_0 = a, \ b_0 = b$$

and the simplified system (also called logistic) obtained by means of the universally valid mean equation

$$b_{n+1} = \frac{1}{2}\left(b_n + \frac{ab}{b_n}\right) \quad \text{with} \quad b_0 = b \quad \text{and} \quad a_n * b_n = ab \ \text{for all} \ n \geq 0$$

are equivalent: they can be transformed into each other and consequently define the same iterated sequences (a_n) and (b_n).

Applications of the Logistic Recursion System

The system of iterated medieties brought to a standardized form

$$\left(x_n * y_n = 1, \ y_{n+1} = \frac{1}{2}\left(y_n + \frac{1}{y_n}\right) \quad \text{with} \quad y_0 = \sqrt{b/a}\right)$$

gives reason to bring another context into play as well. Namely, the remaining recursion formula for the sequence y_n of iterated arithmetic medieties consists in the continuous discrete iteration of the so-called 'logistic function'

$$\varphi(y) = \frac{1}{2}\left(y + \frac{1}{y}\right),$$

which is usefully explained for positive variables $y > 0$. It also plays an important role in **mathematical economic theory** – there it is called the prototype of a '**storage cost function**'. The iteration

$$y_{n+1} = \frac{1}{2}\left(y_n + \frac{1}{y_n}\right)$$

now is exactly the sequence of the continued applications of this function to the prior values:

$$y_0 = \sqrt{b/a}, \ y_1 = \varphi(y_0), \ y_2 = \varphi(y_1) = \varphi(\varphi(y_0)) = \varphi^{(2)}(y_0), \ \ldots,$$
$$y_{n+1} = \varphi(y_n) = \varphi\varphi^{(n)}(y_0) = \varphi^{(n+1)}(y_0).$$

These iterations have been sufficiently studied in discrete dynamics; they lead into the field of **chaos theory,** a new area in the sector of **dynamic systems.**

In the present case, the initial value is $y_0 = \sqrt{b/a} > 1$; therefore, the monotonically decreasing convergence of the iteration sequence is given and it converges to the **fixed point** $y = 1$ of the logistic function – because it is obviously $\varphi(1) = 1$ and therefore $\varphi^{(n)}(1) = 1$, in accordance with the limit properties shown in Theorem 4.3.

Application to the Approximation of the Geometric Mean

We have seen that both sequences $(a_n)_{n \in \mathbb{N}}$ and $(b_n)_{n \in \mathbb{N}}$ of iterated babylonian means approach the geometric mean $z_{\text{geom}} = \sqrt{ab}$ from below and from above. They embed it approximately, just like the inequality chain

$$a = a_0 < a_1 < a_2 < a_3 < \cdots < z_{\text{geom}} < \cdots < b_3 < b_2 < b_1 < b_0 = b$$

once again makes it clear. This inclusion usually leads very quickly to an approximation of the geometric mean, which 'usually' represents an irrational root.

We want to illustrate this with the **example of the octave** in the pythagorean canon

$$6(= a = a_0) - 8(= a_1) - 9(= b_1) - 12(= b_0 = b)$$

Here, the calculator-value corresponds to $\sqrt{ab} = \sqrt{72} = 6\sqrt{2} \approx 8,485281374.$

To start, we already have the first approximation

$$a_1 = 8 < z_{\text{geom}} < 9 = b_1.$$

For the next iterations applies

$$a_2 = \frac{2 * 72}{8 + 9} = \frac{144}{17} \approx 8,470588235 \text{ and } b_2 = \frac{1}{2}(8 + 9) = \frac{17}{2} = 8,5,$$

so that the inequality $a_2 < z_{\text{geom}} < b_2$ has only the error of at most $\pm 3/100$.

$$a_3 = \frac{144}{\frac{144}{17} + \frac{17}{2}} = \frac{144 * 34}{288 + 289} \approx 8,485268631 \text{ and } b_3 = \frac{1}{2}(\frac{144}{17} + \frac{17}{2}) \approx 8,485294118,$$

so that we have already an accuracy up to the 4th – yes almost even up to the 5th – decimal place, like the arrangement

$$a_3 \approx 8,485268631 < z_{geom}(\approx 8,485281374) < 8,485294118 \approx b_3$$

shows. This impressively underlines the nested fast convergence of the two sequences of babylonian medieties. (The accuracy of the data of a_3 and b_3 is given here up to the 8th decimal place!) A further step would even yield the supposed equality

$$a_4 = b_4 \approx 8,485281374 \;(=)\; z_{\text{geom}},$$

Whereby of course the real inequality $a_4 < z_{\text{geom}} < b_4$ still applies; it is only hidden behind the 9th decimal place, i.e. still below the billion digit range.

Application in the Root Calculation

With the help of the babylonian medietary iteration, **roots** can be calculated very fast 'iteratively'; we will show this with a second example for $\sqrt{5}$ (the calculator-value for this root is $\sqrt{5} = 2,236067978$) and we start with this convenient product structure:

$$\sqrt{5} = \sqrt{1 * 5}.$$

We choose $a = a_0 = 1$ and $b = b_0 = 5$ as the start dates. Consequently

$$a_1 = 10/6 = 1.\overline{6} \text{ and } b_1 = 3$$

$$a_2 = \frac{10}{\dfrac{10}{6} + 3} = \frac{60}{28} \approx 2,142857143 \text{ and } b_2 = \frac{1}{2}\left(3 + \frac{10}{6}\right) = \frac{28}{12} = 2,\overline{3}$$

$$a_3 = \frac{10}{\dfrac{60}{28} + \dfrac{28}{12}} = \frac{3360}{1504} \approx 2,234042553 \text{ and } b_3 = \frac{1}{2}\left(\frac{60}{28} + \frac{28}{12}\right) = 2,\overline{238095}.$$

Due to the spread of $a = 1$ and $b = 5$, the convergence is 'slower' than in the example of the octave at $a = 6$ and $b = 12$ – i.e. the determination of $\sqrt{72}$ – but still easily recognizable: Finally, the values differ by only about 1/1000 already in the 3rd step.

Conclusion All in all, these remarks should also show where the 'musical' medieties sequences can be of good service – i.e. also outside of music.

4.4   Bilateral Iterations of Babylonian Medieties

In the constructive-iterative process – to be seen as an '**inner**' iteration – of interlocking babylonian chains of proportions, we have successfully fought our way through a considerable apparatus of formulas with the aim of arriving at the ordering statements of symmetry at a later point, which will then (hopefully), under the bon mot "harmonia perfecta infinita", give us a pleasant overview as well as an understanding of the infinite variety of tones, intervals and relationships.

Moving on to an '**external**' iteration; it is, in a sense, the backward process of the continued generation of new babylonian medieties from the previously obtained data, discussed in the previous section. A basic task that arises, and that gives a good impression of a 'backward' iteration, would be something like this:

Task Given are two data – let's say $a_0 < b_0$. Find magnitudes – written down in advance as a_{-1} and b_{-1} – so that the given data a_0 and b_0 are the harmonic, respectively the arithmetic mean of a_{-1} and b_{-1}. Consequently, we have in any case the arrangement

$$a_{-1} < a_0 < b_0 < b_{-1}.$$

The ascending proportion chain

$$A_{\mathrm{mus}} = a_{-1} : a_0 : b_0 : b_{-1}$$

would then be a typical babylonian musical chain of proportions, as we have studied in the harmonia perfecta maxima babylonica.

This is certainly not too difficult as a task and it should be entrusted at this point to the inquiring minds of our readers. However, one suspects – bearing in mind the efforts made in the seemingly easier reverse task of the last section – that a continuing iterative process of this

Follow-Up Task: *Search for these new data a_{-1} and b_{-1} even newer ones (a_{-2} and b_{-2}) in a way that a_{-1} and b_{-1} are harmonic and arithmetic means of these newer ones*

would take us together into a formula world that would most likely make us despondent; the desire to gain possible results is visibly dwindling – especially since one cannot be sure whether there are any 'results' at all. Or are there?

So we go into the processes of the **bipartite** iterations of babylonian medieties and present a mathematical description – 'from the higher vantage point'. Certainly we must inevitably make a little more use of the mathematical alphabet (the generalizing set language and its basic mathematics) than we have been able to avoid until now '*Less savvy readers please bear with us – but it never hurt to look over the fence*'.

By doing so, we want to advertise for the usefulness of precisely that mathematics which some spirit of the times dismisses as abstract, theoretical or 'dry as dust'. We will solve the given task of bipartite iterations simultaneously and without computational clamps, freely according to the motto:

... that mathematics is not just for calculating – but rather opens its treasure chests to understanding, to ordering and structuring and justifying – as it also wants to serve a generalizing view.

Fig. 4.2 The sectorial
decomposition of the 1st
quadrant of the number plane

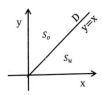

The Iteration of the Babylonian Medieties sequences

As seen, the iteration consists of the continued generation of a harmonic and an arithmetic
mean to the previous data – or in finding those data to whom the given data build the
harmonic, respectively the arithmetic medieties. In the following descriptions we use – in
view of a favorable adaptation to our previous notation symbolism – the basic sets:

$$\mathbb{R}_+^2 = \mathbb{R}_+ \times \mathbb{R}_+ = \{(x,y)|x \geq 0 \text{ and } y \geq 0\}$$

this is the 1st quadrant of the number plane of positive coordinates (including the semi-
axes) and it is the variable space we're working in. Here we consider the two complementary upper and lower sectors,

$$S_o = \{(x,y) \in \mathbb{R}_+^2 | x < y\} \text{ and } S_u = \{(x,y) \in \mathbb{R}_+^2 | x > y\},$$

that are separated by the bisector of the angle (the diagonal $x = y$)

$$D = \{(x,x)|x \geq 0\}$$

The sketch (Fig. 4.2) illustrates the positional relationships.

The following function is adapted to our iteration; it assigns to a pair of data $(a,b) \in \mathbb{R}_+^2$
its two medieties 'harmonic mean' and 'arithmetic mean' – formerly denoted by y_{harm} and
x_{arith}. This is done with slight modification of these notation symbols in the

---

Definition 4.2 (The Operator of Harmonic-Arithmetic Medieties)

The map $\varphi : \mathbb{R}_+^2 \to \mathbb{R}_+^2$, which is given by the rule

$$\varphi(a,b) = (u(a,b), v(a,b)) = (u_{\text{harm}}(a,b), v_{\text{arith}}(a,b)) = \left(\frac{2ab}{a+b}, \frac{a+b}{2}\right)$$

assigns to each pair (a,b) of positive data its two medieties 'harmonic mean' and
'arithmetic mean'

(continued)

Definition 4.2 (continued)
$$u_{\text{harm}}(a,b) \text{ and } v_{\text{arith}}(a,b).$$

Without further ado, we call this mapping the '**babylonian operator**'.

▶ We intend to 'iterate' this operator, which means that we first let it go on two given data (a, b), then take the resulting pair of the new data as new initial data and once again apply the operator to it, and so on. This corresponds exactly to the process we performed as the "one-way (inner) iteration of babylonian medieties" in Sect. 4.3.

In the following theorem – perhaps the most 'mathematical' theorem of our reading – we list all the significant features of the babylonian operator for our purpose. By doing so, we arrange this package of statements in a manner that we will obtain the iteration set at the end.

Theorem 4.4 (Analysis and Functional Analysis of the Babylonian Operator)
The babylonian operator φ has these properties:

1. **Regularity:** As a pair of functions of two variables, φ is continuous in the whole domain of definition, arbitrarily often differentiable everywhere, and analytic.
2. **Symmetry:** In $\mathbb{R}_+^2$ the following symmetry applies: $\varphi(a,b) = \varphi(b,a)$ for all a,b > 0.
3. **Relations about the mapping-domains:**

$$\varphi(S_o) \subseteq S_o, \varphi(D) \subseteq D \text{ and } \varphi(S_u) \subseteq S_o - \text{also } \varphi\left(\mathbb{R}_+^2\right) \subseteq S_o \cup D.$$

When applying the operator φ to any data $a \neq b$, their images $(u, v) = \varphi(a, b)$ are nevertheless ordered; the positional relation $u < v$ always holds.
4. **Fixed point set:** $\varphi(a, b) = (a, b) \Leftrightarrow a = b$,
   and this statement means that exactly the entire diagonal D is a fixed point set: $\varphi_{|D} = \text{Id} -$ that is $\varphi(a, a) = (a, a)$ for all $a > 0$.
5. **Invertibility:** the operator φ is bijective as a mapping $\varphi : S_o \to S_o$ and thus has an inverse operator

$$\psi = \varphi^{-1} : S_o \to S_o.$$

This operator can be specified in a concrete formula and we have the inverse relation:

(continued)

Theorem 4.4 (continued)

$$\varphi(a,b) = (\frac{2ab}{a+b}, \frac{a+b}{2}) = (u,v)$$
$$\Leftrightarrow \psi(u,v) = (v - \sqrt{v^2 - uv}, v + \sqrt{v^2 - uv}) = (a,b).$$

The **operator** ψ, **inverse to** φ, hence computes for two given data $u < v$ exactly the two data $a < b$, for whom u is the harmonic and v the arithmetic mean! Therefore we also have the arrangement $a < u < v < b$.

6. **Hyperbola invariance:** Given $(a,b) \in S_o$, let $\text{Hyp}_{a,\,b}$ be the hyperbola of Archytas within the upper sector S_o through the point (a,b), i.e.

$$\text{Hyp}_{a,b} = \{(x,y) | y = h(x) = \frac{ab}{x}, \ 0 < x < z_{\text{geom}} = \sqrt{ab}\}.$$

The image points of points of this hyperbolic part belong to this hyperbolic part again under the two operators φ and ψ and vice versa. In formulas:

$$(x,y) \in \text{Hyp}_{a,b} \Leftrightarrow \varphi(x,y) \in \text{Hyp}_{a,b}$$
$$\text{and } (u,v) \in \text{Hyp}_{a,b} \Leftrightarrow \psi(u,v) \in \text{Hyp}_{a,b},$$
$$\text{briefly}: \ \varphi(\text{Hyp}_{a,b}) = \text{Hyp}_{a,b} \text{ and } \psi(\text{Hyp}_{a,b}) = \text{Hyp}_{a,b}.$$

In plain language: If x and y are two magnitudes whose product is a $*$ b, i.e. which have the same geometric mean as a and b, then this also applies to the magnitudes $(u,v) = \varphi(x,y)$, whom in turn are by definition the harmonic and arithmetic mean of x and y. All of this is also true the other way round.

7. **The iteration:** Let $(a,b) \in S_o$ be given, then we define for (initially) positive iteration parameters $n = 0, 1, 2, \ldots$ and with the iteration start $(a_0, b_0) = (a,b)$ the sequence of successive babylonian medieties

$$(a_0, b_0) \to (a_1, b_1) = \varphi(a_0, b_0) \to (a_2, b_2) = \varphi(a_1, b_1) = \varphi^{(2)}(a_0, b_0) \to \ldots,$$

this is the general recursion

$$(a_{n+1}, b_{n+1}) = \varphi(a_n, b_n),$$

that can directly be written as the sequence

(continued)

Theorem 4.4 (continued)
$$(a_n, b_n) = \varphi^{(n)}(a_0, b_0), \ n = 0, 1, 2, 3, \ldots$$

where one sets $\varphi^{(0)} = Id$ (the identical mapping). For this iteration sequence follows:

(a) All data (a_n, b_n) lie in S_o on the hyperbola of Archytas $\text{Hyp}_{a, b}$.
(b) The data sequences (a_n, b_n) moves monotonically on the hyperbola $\text{Hyp}_{a, b}$ from (a_0, b_0) towards $(0, \infty)$, which point is also its limit:

$$\lim_{n \to \infty} (a_n, b_n) = (z_{\text{geom}}(a, b), z_{\text{geom}}(a, b)).$$

8. **Backward iteration:** If we set $n = -m$ with positive m in the case of negative iteration parameters, an iteration can also be done 'backwards' by setting

$$\varphi^{(n)} = \varphi^{(-m)} = (\varphi^{-1})^{(m)} = \psi^{(m)}.$$

Therefore, the backward iteration sequence $\varphi^{(-m)}$, $(m = 0, 1, 2, \ldots)$ corresponds to the forward iteration of ψ. For the sequence of iteration points

$$(a_{-m}, b_{-m}) = \psi^{(m)}(a_0, b_0), \ m = 0, 1, 2, 3, \ldots$$

two characteristic statements now also apply:

(a) All data (a_{-m}, b_{-m}) lie in S_o on the hyperbola of Archytas $\text{Hyp}_{a, b}$.
(b) The data sequences (a_{-m}, b_{-m}) moves monotonically on the hyperbola $\text{Hyp}_{a, b}$ from (a_0, b_0) in the northward direction to the pole point (the 'north pole'$(0, \infty)$), thus in the opposite direction to the iteration points of the forward iteration of the operator φ. For the limit process applies:

$$\lim_{m \to \infty} (a_{-m}, b_{-m}) = (0, \infty).$$

9. **Bipartite iteration series:** The bipartite series

$$(a_n, b_n) = \varphi^{(n)}(a, b), \ n = 0, \pm 1, \pm 2, \pm 3, \ldots$$

(continued)

Theorem 4.4 (continued)

lies on the hyperbola of Archytas with starting values (a, b). The course of the magnitudes of the sequences (a_n) and (b_n) happens as follows:

(a) The sequence $(a_n)_{n \in \mathbb{Z}}$ grows monotonically with left-hand side limit 0 and right-hand side limit $z_{geom}(a, b)$.

(b) The sequence $(b_n)_{n \in \mathbb{Z}}$ falls monotonically with left-hand side limit ∞ and right-hand side limit $z_{geom}(a, b)$.

Conclusion: The following sequences of magnitudes have been created, which – provided with positive indices (n) – is arranged as follows:

$$0 < \cdots a_{-n} < \cdots < a_0 < \cdots < a_n < \ldots z_{geom} \ldots < b_n < \cdots < b_0 < \cdots < b_{-n} \ldots < \infty$$

The entire sequence strives inwardly from right and left towards z_{geom} and outwardly towards 0 respectively grows unrestricted. All equal indexed pairs of magnitudes (a_n, b_n) have the central geometric mean z_{geom} as geometric center.

The proof actually consists 'only' in a summarizing observation of earlier results. We skip point (1) because it belongs to the analytic common knowledge. It is also clear that the variables a and b *can* be interchanged, This can be seen directly from the formulas. Statement (3) is helped by the inequality of the medieties: The harmonic mean – for different data a and b – is always smaller than the arithmetic mean -, so all sets of values for both data from S_o and data from S_u belong to S_o. On the diagonal D, a and b are equal and identical with all their mean values – which correlates simultaneously with statement (4).

The Invertibility (5) is obviously only possible in the upper sector (including the diagonals); but there the proof consists simply in seeing that we can solve the equation

(A) $2ab/(a + b) = u$ and $(a + b)/2 = v$

to given data u, v of the upper sector S_o uniquely by data a and b – also from the upper sector S_o. This happens by clever handling. By simple inversion of the second equation $2/(a + b) = 1/v$, the system (A) is equivalent to the system

(B) $uv = ab$ and $a + b = 2v$,

from whom, by the way, we also see that the solutions (a, b) lie on the same hyperbola as (u, v) since their product is equal to the product of u and v. We substitute in (B) from the second equation the term $a = 2v - b$ into the first equation and obtain a quadratic equation for b, so that the system (B) is equivalent to the system

(C) $b^2 - 2vb = -uv$ and $a = 2v - b$

The solutions of this quadratic equation are

$$b = v \pm \sqrt{v^2 - uv},$$

and therefore we find the values

$$a = v \mp \sqrt{v^2 - uv}.$$

These two solution pairs are the solutions from S_u and S_o mirrored at the diagonal D and the unique solution from the upper sector S_o, for which must be $a < b$, is finally the data pair

$$a = v - \sqrt{v^2 - uv}, \quad b = v + \sqrt{v^2 - uv}.$$

Moving on to statement (6) that we have just shown for the inverse function ψ, namely by establishing the equality of the products uv and ab. A *simple principle of set mathematics* serves this purpose:

Principle: *"If the hyperbola is invariant under the operator ψ, then also under its inverse function φ".*

Incidentally, the theorem of Iamblichos/Nicomachus could also be quoted here, according to which harmonic and arithmetic means have the same geometric medieties – one of the striking facts of the harmonia perfecta maxima.

Now both statements (a) of (7) and (8) are a direct consequence of the information from (6) that we bring to bear on each iteration step The monotonicity always results from the fundamental ordering of the two medieties to their magnitudes $x < y$,

$$x < \text{harmonic mean of} (x, y) < \text{arithmetic mean of} (x, y) < y.$$

What about the limit values? Obviously we lack a concrete formula of the two magnitude sequences – i.e. a formula that depends on the sequence parameter n. It would probably be very complicated, and the success would be doubtful.

In accordance with the wise insight that ***the strength of mathematics*** is ***its theory,*** of which it has been known since Aristotle that it ***is the highest form of practice***, we want to take this to heart and present a little cabinet piece of fiddly analysis, which trickily and without calculation leads us to the desired results:

First of all, it is sufficient to show that the left-handed (monotonically decreasing) sequence of magnitudes a_{-n} converges to 0 and that the right-handed (monotonically decreasing) sequence of magnitudes b_n converges to the geometric mediety z_{geom} – because via the product equation

$$a_n b_n = ab = z_{geom}^{\ 2}$$

all statements for one sereis can be transferred to the other accordingly.

Since both positive series are monotonically decreasing, they each have a limit according to **the** famous **completeness principle of Bolzano-Weierstrass** – say

$$a_{-n} \to \alpha \geq 0 \text{ (for } n \to \infty) \text{ and } b_n \to \beta \geq z_{geom} \text{ (for } n \to \infty).$$

Let us assume $\alpha > 0$ – that is, the opposite of what we want to show to be true, namely that $\alpha = 0$. Then b_{-n} also converges to ab/α. Now we have the iteration

$$P_n = (a_{-n}, b_{-n}) = \psi(a_{-n+1}, b_{-n+1}) = : \psi(P_{n-1})$$

By assumption, the series of points P_n on the hyperbola converges to the finite limit point $P_\infty = (\alpha, ab/\alpha)$ – and now comes the **tricky crucial argument:**

Principle*: Continuous functions transfer convergent sequences into convergent sequences: The image sequence of a convergent sequence is convergent if the operator is continuous!*

Therefore, we observe the convergence

$$P_\infty \leftarrow P_n = (a_{-n}, b_{-n}) = \psi(a_{-n+1}, b_{-n+1}) = : \psi(P_{n-1}) \to \psi(P_\infty).$$

Hence, P_∞ is a fixed point of ψ. But as we know, all fixed points of φ and thus also those of the inverse ψ, lie on the diagonal D. However, then

$$\alpha = ab/\alpha = z_{geom}$$

had to be true – a blatant contradiction to the fact that the sequence of data a_{-n} moves monotonically decreasing away from z_{geom} towards the zero point. So our assumption is wrong, and it is $\alpha = 0$.

For the sequence b_n with positive indices we argue completely analogously: Assuming that this series does not converge to z_{geom} would result in a limit β greater than z_{geom} due to monotonicity. But then we have again a convergent sequence of points Q_n on the hyperbola, which this time suffices the recursion

$$Q_n := \varphi(Q_{n-1})$$

The continuity of φ leads this process into a fixed point equation, so that the limit point Q_∞ would be a fixed point of φ – but it is not, since it can lie on the hyberbola, but not on the diagonal – only the point (z_{geom}, z_{geom}) of the geometric mean fulfills this.

Finally, the last point (9) is a summary of the preceding series mappings, proving this (long) theorem.

> Quote from my math teacher from my school days:
>   "Thinking $\times$ calculating $=$ constant".

Now let us return to the question posed at the beginning of this section, namely, how to get to the magnitudes a and b with $a < b$ for two given data u and v with $u < v$, to whom the given data (u) is the harmonic mean and (v) is the arithmetic mean. In other words the

Problem: *For a given proportion $u : v$, find those magnitudes a and b such that the chain $a : u : v : b$ is a babylonian (musical) chain of proportions.*

The general **answer** is provided by our formula in part (5) of the theorem: by applying the inverse of the babylonian operator to the given data (u, v), we use the formulas

$$(a, b) = \psi(u, v) = \left(v - \sqrt{v^2 - uv}, v + \sqrt{v^2 - uv} \right).$$

These formulas can also be written using the **reciprocal frequency measure** $p = u/v$ of the given proportion $u : v$ as follows

$$(a, b) = \psi(u, v) = (v - v\sqrt{1 - p}, v + v\sqrt{1 - p})$$

and therefore represent the desired response to the task.

Specifically, in the particularly interesting case of so-called '**simply superparticular**' intervals – that is, intervals that correspond to the special proportions of the form $(m - 1) : m$ (such as octave $1 : 2$, fifth $2 : 3$ and so on) – this gives rise to the remarkable formula

$$(a, b) = \psi(u, v) = \left(v - \sqrt{v}, v + \sqrt{v} \right) = \left(m - \sqrt{m}, m + \sqrt{m} \right).$$

This formula then leads in a flash (and also by mental arithmetics) to the sought magnitudes a and b – while at the same time it is clear, that this leads exactly in the case of square numbers (m) to a rational (and at the same time integer) result.

In the final example block, we will give some cases systematically:

Example 4.2 The Backward Iteration Process

For the cases where in the proportion u : v ≅ (m−1) : m the number m is a square number, the calculated babylonian proportion chains $a : u : v : b$ are as follows:

$u : v$	Interval	$a : u : v : b$	$a:b$	Interval
3 : 4	Fourth	2 : 3 : 4 : 6	1 : 3	Duodecime, twelfth
8 : 9	Large whole tone	6 : 8 : 9 : 12	1 : 2	Octave
15 : 16	Diatonic semitone	12 : 15 : 16 : 20	3 : 5	major sixth
24 : 25	Small chroma	20 : 24 : 25 : 30	2 : 3	*Fifth*
35 : 36		30 : 35 : 36 : 42	5 : 7	*Ecmelian third*
48 : 49	Septimal diësis	42 : 56 ≅ 3 : 4	3 : 4	*Fourth*
80 : 81	Syntonic comma	72 : 80 : 81 : 90	4 : 5	Major third

Regarding the last example, so all you have to do is subtract and add the root (9) from v (in the example: the root of 81) and then you receive the outer proportion-numbers (72) as well (90) – hence you have the final babylonian chain of proportions.

We can even take a twofold backward iteration out of these foregoing examples: For this purpose we form to the outer magnitudes of the first iteration of the proportion 48 : 49 – i.e. from 42 : 56 ≅ 3 : 4 – whose iteration corresponds to

$$2 : 3 : 4 : 6 \cong 28 : 42 : 56 : 84$$

and then add the inner proportion 48 : 49. We thus obtain the sequence

48 : 49 →	42 : 48 : 49 : 56 →	28 : 42 : 48 : 49 : 56 : 84

of an iterative process.

A further step, however, would lead us out of the rationality of numbers since one can simply calculate with the initial formula because of 28 : 84 = 1 : 3

$$(a, b) = \psi(1, 3) = (3 - 3\sqrt{2/3},\ 3 + 3\sqrt{2/3}) \approx (0, 55...,\ 5, 45...)$$

and obtain these rounded values of irrationalities. ◀

These examples lead – backwards – to some of the models we calculated earlier as babylonian chains.

4.5   The Harmonia Perfecta Infinita: The Infinite Harmony

In various recursive procedures, we have created sequences of mean value magnitudes that can be composed to 'infinite-step' proportion chains. These magnitudes are nested constructs of babylonian medieties – primarily of the arithmetic-harmonic genre.

If the simple chain of the pythagorean canon $6 : 8 : 9 : 12$ as well as that of the extended diatonic canon $6 : 7, 2 : 8 : 9 : 10 : 12$ had already astonished the ancient world by their inner symmetries – i.e. proportional ratios – and caused them to immortalize these small numerical miracles as a **harmonia perfecta maxima:** How much more would the belief in mathematical-musical miracles have increased if it had turned out that this harmonia perfecta maxima is at home in a thousandfold way in the **infinitely many sub-chains of these unlimited chains of proportions** and provides a – one would have said earlier: *divine* – order in the system of musical proportions and their intervals: the harmonia perfecta maxima thus becomes the ***harmonia perfecta infinita.***

(A) The sequence of the contra-medieties
Firstly we consider the contra-medieties (x_n) from Sect. 4.2. The magnitudes x_n are exactly the medieties to the division parameter $q^n = (b/a)^n$. Therefore the sequence of proportions

$$(x_n - a) : (b - x_n) \cong b^n : a^n \quad \text{for} \ \ n = 0, \pm 1, \pm 2, \ldots$$

is realized by the data x_n that lie monotonically ordered in the interval $[a, b]$, which can be represented in the following formula:

$$a < \cdots x_{-n} : \ldots : x_{-2} : x_{-1} : x_0 : x_1 : x_2 : \ldots : x_n : \ldots < b \quad (n = 0, 1, 2, \ldots).$$

Subsequently, the concrete magnitude formulas

$$x_n = \frac{a^{n+1} + b^{n+1}}{a^n + b^n} = ab \frac{a^{-(n+1)} + b^{-(n+1)}}{a^{-n} + b^{-n}}, \ \ \text{with} \ \ n = 0, \pm 1, \pm 2, \ldots$$

are obtained. Now we're wondering:

Question: What symmetries might this – towards its edges a and b monotonously converging – infinite monster of proportions possess?

The **answer** is given by the following theorem:

Theorem 4.5 (Harmonia Perfecta Infinita Contra-Babylonica)
The sequence of contra-medieties $(x_n)_n \in \mathbb{Z}$ defines a bipartite unrestricted continued proportion chain C_{mus}^{∞}, which we write down with positive indices (n) like this:

$$C_{mus}^{\infty} = \ldots x_{-n-1} : x_{-n} : \cdots : x_{-2} : x_{-1} : x_0 : x_1 : x_2 : \cdots : x_n : x_{n+1} : \ldots$$

The magnitudes x_n all lie in the real interval $[a, b]$, and by Theorem 4.2, the magnitudes converge monotonically from the inside to the edges a and b of the initial proportion $a : b$. The following regularities can be shown:

1. **Global symmetry and center property**
   C_{mus}^{∞} is **globally symmetric,** meaning that C_{mus}^{∞} is similar to its reciprocal - in formula

$$C_{mus}^{\infty} \cong \left(C_{mus}^{\infty}\right)^{rez}.$$

Hereby the geometric mean

$$z_{geom} = z_{geom}(a, b) = x_{-(1/2)}$$

is the center of symmetry of the entire chain C_{mus}^{∞} with respect to whom the global symmetry is also measured: The magnitudes x_n and $x_{-(n+1)}$ are mirrored with respect to the center z_{geom} as well as to the proportion $a : b$ – consequently

$$x_n^* = x_{-(n+1)} \quad \text{(for all } n = 0, \pm 1, \pm 2, \ldots);$$

and we have the proportions/equations

$$x_n : b \cong a : x_{-(n+1)}$$

$$\Leftrightarrow x_n * x_{-(n+1)} = ab = z_{geom}^2 \quad \text{(for all } n = 0, \pm 1, \pm 2, \ldots).$$

For example: In the special case n = 0 we see that the magnitudes

$$x_0 = x_{arith}(a, b) \text{ and } x_{-1} = y_{harm}(a, b), \quad x_1 = y_{co-harm}(a, b) \text{ and } x_{-2} = x_{co-arith}(a, b)$$

are mirrored – as in the harmonia perfecta maxima diatonica.

(continued)

Theorem 4.5 (continued)

2. **The system of babylonian subproportion chains**

   The following ordered systems of babylonian subproportion chains of the proportion chain C_{mus}^{∞} exists:

   (A) Any 2-step subproportion chain of the form

   $$G_n = x_{-(n+1)} : x_{-(1/2)} : x_n \quad (\text{with } n = 0, 1, 2, \ldots)$$

   is **geometric**; all chains G_n have the same geometric center

   $$x_{-(1/2)} = z_{\text{geom}}(a, b).$$

   (B) Any 2-step subproportion chain of the form

   $$A_n = x_{-n} : x_0 : x_n \quad (\text{with } n = 0, 1, 2, \ldots)$$

   is **arithmetic**; all chains A_n have the same arithmetic mean

   $$x_0 = x_{\text{arith}}(a, b).$$

   (C) Any 2-step subproportion chain of the form

   $$H_n = x_{-1-n} : x_{-1} : x_{-1+n} \quad (\text{with } n = 0, 1, 2, \ldots)$$

   is **harmonic**; all chains H_n have the same harmonic mean

   $$x_{-1} = y_{\text{harm}}(a, b).$$

3. **Symmetries of reciprocity**

   For any integer proportion index n, the **arithmetic chain A_n is reciprocal to the harmonic chain H_n** – in formulas:

   $$H_n = A_n^{\text{rez}} \quad \text{and} \quad A_n = H_n^{\text{rez}} \quad \text{for all } n \in \mathbb{Z}.$$

Special case: In the case of the octave canon $a : b \cong 6 : 12$ the 5-step subproportion chain $D_{\text{mus}}(6, 12)$ of C_{mus}^{∞} which is complemented by the outer magnitudes a and b, is the chain

(continued)

Theorem 4.5 (continued)

$$D_{\mathrm{mus}}(6,12) = a : x_{-2} : x_{-1} : x_0 : x_1 : b = 6 : 7, 2 : 8 : 9 : 10 : 12,$$

and therefore this is exactly the proportion chain of **the** complete musical **canon of pure diatonicism** as discussed in Sect. 3.6. It consists entirely of mirrored magnitude pairs (a and b, x_{-2} and x_1, x_{-1} and x_0).

Proof According to Theorem 4.2 in Sect. 4.2, in which we developed a formula world of the sequence of medieties (x_n), the given subproportion chains G_n are geometric, A_n arithmetic and H_n harmonic due to the symmetry formulas (3).

Since all geometric chains G_n always have the same geometric mean $z_{\mathrm{geom}}(a, b)$ as their own geometric mean, it is also the center of the whole chain C_{mus}^∞. And by applying the symmetry principles from Theorems 3.6 and 3.7 – as we have done many times before – the global symmetry of the whole infinite proportion chain C_{mus}^∞ follows immediately. The mirror property of the magnitudes x_n and $x_{-(n+1)}$ follows as a further special case from the center property of the common center of all chains G_n – this is just the concrete version of the global symmetry of the whole chain C_{mus}^∞. Thus, statements (1) and (2) are clear. What remains is the proof of the reciprocity relation (3):

To (3): We check the equivalences of the proportions

$$x_n : b \cong a : x_{-(1+n)} \quad \text{and} \quad a : x_{-n} \cong x_{-1+n} : b$$

If we now had the, perhaps obvious, idea of using the obtained concrete magnitude formulas for these magnitudes x_k, we would however face an extensive calculation with questionable success. Nevertheless, everything is simple: Crosswise multiplication leads the proportions into the equations

$$x_0 * x_{-1} = x_{-(1+n)} * x_n \quad \text{and} \quad x_{-n} * x_{-1+n} = x_{-1} * x_0$$

However, all four products are equal – namely with the product $a*b$ -, because the factors are mirrored magnitudes, as we just read in Theorem 4.2 (3) (for the product $x_{-n}*x_{-1+n}$ one sets the index $(-n)$ for the index n in the mirror equations (3)).

An In-Between Remark

We could have reached our target just as well with the cross rule and the exchange rule since after all

$$x_0 : 1 \cong ab : x_{-1},$$

applies with whom one would have come from the proportions to be examined to the reflections

$$x_n : b \cong a : x_{-(1+n)} \quad \text{and} \quad a : x_{-n} \cong x_{-1+n} : b.$$

This also proves Theorem 4.5 about the harmonia perfecta infinita for this case of the sequence of contra-medieties.

(B) The sequence of arithmetic-harmonic medieties

Next, we consider the bipartite series $(a_n, b_n)_n \in \mathbb{Z}$ of all iterated arithmetic-harmonic medieties of a given starting proportion $a_0 : b_0 = a : b$, which we compiled in Sect. 4.4 using the **babylonian operator** φ.

With the help of this operator, a double-bipartite – inward and outward – sequence of iterated arithmetic and harmonic medieties

$$0 < \cdots a_{-n} < \cdots < a_0 < \cdots < a_n < \cdots z_{\text{geom}} \ldots < b_n < \cdots < b_0 < \cdots < b_{-n} \ldots < \infty$$

can be built and joined together to form a single (enormous) chain of proportions that ultimately has the character of unrestrictedly progressing series of proportions at four ends:

Towards the inside both sequences strive monotonously to the geometric center,

$$\lim_{n \to \infty} (a_n) = z_{\text{geom}}(a, b) \quad \text{and} \quad \lim_{n \to \infty} (b_n) = z_{\text{geom}}(a, b),$$

and outwardly one strives to 0, the other to ∞,

$$\lim_{m \to \infty} (a_{-m}) = 0 \quad \text{and} \quad \lim_{m \to \infty} (b_{-m}) = +\infty.$$

This is how the fourfold infinitely continuing chain of proportions comes into being, which we can also write down as a gigantic musical chain of proportions with a single positive indexing

$$B_{\text{mus}}^{\infty} = \cdots a_{-n} : \ldots : a_0 : \ldots : a_n : \ldots : \left(z_{geom} \right) : \ldots b_n : \ldots : b_0 : \ldots : b_{-n}$$
$$: \ldots (n = 0, 1, 2 \ldots)$$

This bipartite proportion chain obviously contains the unipartite one considered in the earlier Sect. 4.3 'as a subproportion chain'; it runs from the initial magnitudes (a, b) inwards, which is why we also use the same symbol. But then the unifying notation for the closed representation is helpful:

$$B^\infty_{mus} = \ldots a_n : a_{n+1} : \cdots : b_{n+1} : b_n : \ldots \text{(with } n \in \mathbb{Z} \text{ (that is } n = 0, \pm 1, \pm 2, \pm \ldots)).$$

For usage that only requires forward iteration, simply suppress all negative indices; replace the large index range $\mathbb{Z}$ with $\mathbb{N}_0 = 0, 1, 2, \ldots$.

Question: *"What ordered structures does this fourfold unrestrictedly continued chain of proportions possess? – And: Can such a thing exist at all?"*

The following theorem gives us the positive **answer:**

Theorem 4.6 (The Harmonia Perfecta Infinita of the Babylonian Medieties Series)

1. **Global symmetry and center property**

   The musical chain of proportions B^∞_{mus} is globally symmetrical – i.e.

   $$\left(B^\infty_{mus}\right)^{rez} = B^\infty_{mus},$$

   and it has the geometric mean $z_{geom}(a, b)$ as its center of symmetry.

   It consists of an infinite series of **concentrically arranged,** nested, 2-step geometric proportion chains G_n with a common center $z_{geom}(a, b)$,

   $$G_n = a_n : z_{geom} : b_n \quad \text{(for all } n = 0, \pm 1, \pm 2, \ldots).$$

   These 2-step chains are built from the babylonian medieties trinity in the structure "harmonic – geometric – arithmetic".

2. **The mirroring proportions and equations**

   Equal-indexed magnitudes a_n and b_n are mirrored to each other with respect to the center $z_{geom}(a, b)$ and to any other choice of a pair of magnitudes (a_m, b_m) : For all indices n *applies*

   $$b_n = a_n^* \quad \text{respectively} \quad a_n = b_n^*.$$

   For all indices $n, m \in \mathbb{Z}$ we have the proportions

   $$a_m : b_n \cong a_n : b_m \quad \text{and} \quad a_m : a_n \cong b_n : b_m,$$

   and especially the proportions to the reference data (a, b) are mirrored:

   $$a : a_n \cong b_n : b \quad \text{and} \quad a : b_n \cong a_n : b \quad \text{(for all } n = 0, \pm 1, \pm 2, \ldots)$$

(continued)

Theorem 4.6 (continued)

3. **Systems of Babylonian subproportion chains**

    Each 3-step subproportion chain P_n of B_{mus}^∞, arranged concentrically, of the form

$$P_n = a_n : a_{n+1} : b_{n+1} : b_n = a_n : y_{harm}(a_n, b_n) : x_{arith}(a_n, b_n) : b_n$$

is again a babylonian, respectively pythagorean proportion chain of the type of the harmonia perfecta maxima shown in Theorem 3.2 and therefore satisfies both theorems of Nicomachus and Iamblichos. For the anterior subproportion chain H_n and the posterior subproportion chain A_n of P_n

$$H_n = a_n : a_{n+1} : b_{n+1} \text{ and } A_n = a_{n+1} : b_{n+1} : b_n$$

then consequently applies

1. H_n is harmonic, and A_n is arithmetic,
2. H_n and A_n are reciprocal to each other,

$$H_n = A_n^{rez} \text{ and } A_n = H_n^{rez},$$

from which finally results a wide-spreading net of **infinite internal symmetries of proportions** – since this is valid for all $n = 0, \pm 1, \pm 2, \ldots$

Special case: We note that the chain of the babylonian (pythagorean) octave canon in the case of given magnitudes $a = a_0 = 6$ and $b = b_0 = 12$ is the sub-chain

$$P_0 = a_0 : a_1 : b_1 : b_0 = 6 : 8 : 9 : 12$$

of B_{mus}^∞ and is thus a component and starting point of the iteration process.

The proof consists, on the one hand, in the direct application of the center principle for the geometric mediety from Theorem 3.6, and, on the other hand, each sub-chain

$$P_n = a_n : a_{n+1} : b_{n+1} : b_n$$

is indeed constructed in such a way that the principle of harmonia perfecta maxima in the simple babylonian variant – that is Nicomachus' Theorem 3.2 for the 3-step harmonic-arithmetic (read: Babylonian) medieties chain

$$P_{\mathrm{mus}} = a : y_{\mathrm{harm}} : x_{\mathrm{arith}} : b$$

is applied. All statements of this basic situation together with the detailed descriptions of Theorem 4.4 lead to all essential core statements of the theorem.

Concluding Remarks   At this point we have to confess something to our readers: We have allowed ourselves to be seduced a little by the temptations to which mathematicians are always subject:

Joy and sorrow: Once a question pops up, there is no rest until it has been explored in its ramified details, disassembled and reassembled, thought through anew – and finally understood. Unfortunately, however: There is said to be the bon mot by Goethe that a problem, which one would entrust to a mathematician for solution, returns afterwards – and proudly solved in its entirety – as a completely unknown being. . .

Don't worry: we already know, for example, that a babylonian chain of proportions

$$P_n = a_n : a_{n+1} : b_{n+1} : b_n,$$

which began at the start data 20 : 30 of a pure fifth, first comes up with the melodious chain

$$a_0 : a_1 : b_1 : b_0 \cong 20 : 24 : 25 : 30$$

It contains after all the beautiful major chord 20 : 25 : 30 and the no less beautiful minor chord 20 : 24 : 30. In the next iteration, however, the *quarter-tone interval 'small chroma'* **24 : 25** is already *divided arithmetically and harmonically, and microtones arise.*

After only a handful of iterated averages, all the tones of the sub-chain P_n have moved so close together that their frequencies differ only in remote decimal regions. There is no question of chords, and possible beats have frequencies of perhaps several decades, one could easily calculate.

No, we have allowed ourselves to be tempted to pursue the desire to comb through a universal harmonia to its end. In this wish, however, we are fulfilling a monstrous doctrine of the great mathematical scholar Carl-Gustav-Jacob **Jacobi** (1804–1851 in Potsdam, Germany),

„. . .Mathematics can only be called such if it is devoid of any application. . .',

and hold the same steadfastly against the spirit of the times.

The Music of Proportions

5

> ...One is only truly a musician who gains his knowledge of
> music-making in deliberative reasoning and not from practical
> experience but from the compulsion to think.... *Anitius Manlius
> Severinus Boethius (From [5], p. 126)*

In this chapter we will turn explicitly to musical matters and bring the concepts stemming from the ancient mathematically motivated theory of proportions to bear on the areas of music theory that concern them. We will first give an initial overview:

In the **first** Sect. 5.1, we again clearly present the relation between tones and intervals on the one hand and proportions on the other, in particular by using the concept of the **monochord** to connect *'non-measurable' proportions (musical intervals) by 'measurable' proportions (geometric lengths)* and to be able to describe the building of musical pitch, interval and scale structures with this effective tool. In addition, we present the most important methods for using the possibilities of the mathematics of proportions to obtain a **systematics** for generating musical interval and tone systems.

In the **second** Sect. 5.2 we will deal with concrete cases: Three systems and their most important interval structures – including their **semitonia and commas** – will be presented in great detail: These are in detail

- the pythagorean,
- the just diatonic,
- and an ecmelic system.

The **third** Sect. 5.3 offers an insight into playing with **chords** and their chains of proportions: The **questions**

K. Schüffler, *Proportions and Their Music*,
https://doi.org/10.1007/978-3-662-65336-4_5

- *How do major and minor become visible in the numerical realm?*
- *What symmetries of their proportions are characteristic?*

are certainly only a beginning of analyses in this regard.

In the **fourth** Sect. 5.4, we connect the geometric chains of proportions with structures of **microtonal intervals** thus gaining a methodologically guided overview of some relations such as the discussion of the **questions:**

- *How are the semitonia of the just diatonic related to each other?*
- *Are there parallels to medietary properties?*

The **fifth** Sect. 5.5 takes us into the world of **ancient greek tetrachords.** We present their classification by gender and family, followed by calculations based on proportion theory

- **dorian, phrygian** *and* **lydian** *tetrachords*
- *in the* **diatonic, chromatic** *or* **enharmonic** *genders.*

But also the tetrachord for the **'musical proportion of Iamblichos'** will cause surprises with its peculiar gradations. And the questions of how the sometimes bizarre interval *relations* (such as an interval proportion 48 : 49) of greek tetrachordics can be explained could certainly be discussed from these new points of view.

In the **sixth** Sect. 5.6, we will have a little insight into the **ecclesiastical** and **gregorian modes.** Starting from the well-known greek-antique universal tonal system, the **'systema teleion'**, we develop the so-called **octochord structures** of these modes – sometimes called **'oktoechos'** – whereby we develop and present three methods for this:

- the combination method,
- the octochord method,
- the step method.

In the **last** Sect. 5.7, we then take a closer look at the **mathematics of proportions of the organ.**

▶   In no other instrument than the organ has the reference to proportions been preserved so distinctly. The understanding of the laws of pitch of the stops dispositions as well as the sound-physical interaction of individual organ stops obeys – ultimately – the laws of the theory of proportions.

The rules of monochordics and the theory of proportions are therefore also steady companions of our reading. Several examples from organ practice are then explained from **groups** of stops (such as mixtures and cornets), mostly composed of various **aliquots,** which after all correspond mathematically to our **chains of proportions.**

5.1 From Monochordium to Theory: Musical Intervals and Their Proportional Calculus

If we trace the path that the theory of musical concepts has taken from antiquity, we see that it has been accompanied at every turn by models -whether abstract or concrete. In terms of tonal systems, their intervals, their construction into scales and chords, the monochord is this model above all others:

Monochord Model *"You have a taut string and you study the tones that are produced when the string is divided in various ways".*

In principle, a sounding reed (flute) could also be used for this purpose, but it is obviously impractical; the organ's construction of geometrically scaled and dimensioned pipes may support this aspect. We will return to this in more detail in the last Sect. 5.7 of this chapter.

Musical intervals are described by proportions, and in this section we imagine, where it seems possible, an accompanying model realizing these proportions by means of a monochord – if necessary by multiple ones of these – translated into tones and music.

Thus, in this section we're going to look at

- the regularities between intervals and proportions on the monochord
- and the five traditional principles for the architecture of sound systems

First, however – forgiveness – we must remain somewhat theoretical and briefly describe once again the fundamental relation between music and mathematics – here: between musical intervals and mathematical proportions – perhaps for the umpteenth time – and in doing so clearly highlight the most important parameter – the frequency and the proportion measure – in their significant musical handling. Incidentally, we refer in the context of this discussion to the handling of the proportions for the intervals representing them (keyword: "octave 1 : 2 instead of 2 : 1").

Definition 5.1 (Musical Intervals and Their Measures)

If a and b *are* two positive (natural – but also any real -) numbers, the musical interval $[a, b]$ is defined in mathematical language by

$$[a, b] = \text{set of all ordered pairs of tones } \left(\tilde{a}, \tilde{b} \right) \text{ for which } \tilde{b}/\tilde{a} = b/a.$$

Thus $[a, b]$ corresponds to the totality of all proportions $\tilde{a} : \tilde{b}$ similar to the proportion $A = a : b$, and we can additionally say

---

(continued)

Definition 5.1 (continued)
$$[a,b] \equiv \text{class of all proportions } \widetilde{A} = \widetilde{a} : \widetilde{b} \text{ with } \widetilde{a} : \widetilde{b} \cong a : b.$$

In music theory, there are three main measures for intervals $I = [a,b]$:

1. The **frequency measure of I** is the quotient $|I| = b/a$.
   This quotient is by definition independent of the chosen representative of the whole class $[a, b]$. Thus applies the mathematically written characterization

$$[a,b] = [c,d] \Leftrightarrow b/a = d/c \Leftrightarrow A = a : b \cong C = c : d.$$

   Two intervals are therefore equal if they have the same frequency measure – or if their descriptive proportions are similar.

2. The **proportion measure** is – basically – the description of an interval as a proportion $a : b$, whereby one strives for a form of the magnitudes that is as shortened as possible and, if possible, integer.
   The proportion measure translates –in general – verbal interval indications into the language of proportions; via the frequency or cent measure, a numerical description is then made, if required, together with its use.

3. The **cent measure** is the logarithmic form of the frequency measure - adjusted to the octave,

$$\text{ct}(I) = 1200 * \log_2(|I|) = 1200 * \log_2\left(\frac{b}{a}\right) = 1200 \frac{\ln|I|}{\ln 2}$$

   and for example we have specially the measure: ct (Octave $1 : 2$) $= 1200\,\text{ct}$ (read 'cent' for ct).
   The **importance of the cent measure** lies primarily in its metric property:
   The cent measure of the **sum** (adjunction, stratification) of two intervals is also the **sum** of the measures; in the case of the frequency measure, this is the product – which is significantly less useful with scales and multiple interval additions.

Hence, the proportion measure of a just fifth is the mere proportion $2 : 3$, the frequency measure is the fraction $3/2$ – or its numerical value 1.5; the cent measure ct(Fifth) is irrational and has the rounded value 701.95 ct.

If we have a concrete pair of tones $\left(\widetilde{a}, \widetilde{b}\right)$ of an interval class $[a, b]$, then these concrete numbers $\left(\widetilde{a}, \widetilde{b}\right)$ are to be understood as front, respectively back proportional and simultaneously also as fundamental frequencies, whereby they can be interpreted as physical tones. This concrete frequency meaning gets of course lost in the entire class: an interval is thus not bound to the real pitch value of its two tones but exclusively to their numerical proportion.

Thus $[1, 2]$ is the interval of an octave – no matter which realizing tones are chosen for it. Because the realizing magnitudes a, b are free – as long as the quotients b/a *steadily* yield the same value -, all objects

$$[1, 2], [2, 4], [311, 622], [440, 880], [2024, 4048]$$

are always 'the octave'; and likewise, the whole-tone steps $[8, 9]$ and $[16, 18]$ would be the same interval (namely, a major pythagorean whole tone [tonos]).

Conclusion: The notations [a, b] and a : b are only symbolically different, considering their respective identifications within their similarity classes. Nevertheless, the argumentative or computational handling can sometimes be affected by this.

Now, we come to the process to adjoin two or more intervals to a new interval: their **sum**. Just as we have fused (= multiplied $\odot$) several proportions into a new **proportion**, several intervals are also layered (= adjunct $\oplus$) into a new **interval** and then these processes $\oplus$ and $\odot$ correspond to each other; the mathematical specifications are namely: For two intervals $I_1 = [a_1, b_1]$ and $I_2 = [a_2, b_2]$ we define the construct

$$I_1 \oplus I_2 = [a_1, b_1] \oplus [a_2, b_2] = [a_1 a_2, b_1 b_2] \equiv \text{the sum of the intervals } I_1 \text{ and } I_2$$

which is the musical interval of **adjunction** in the form of an **addition** or **sum** of both intervals, and its proportion class corresponds exactly to the **fusion of** the proportions $A_1 = a_1 : b_1$ and $A_2 = a_2 : b_2$, which are indeed given by

$$A_1 \odot A_2 = (a_1 : b_1) \odot (a_2 : b_2) \cong (a_1 a_2 : b_1 b_2)$$

and regarded as their 'product', see Sect. 1.4. We will remarke: In capters 2 and 3 we have used the symbol $\oplus$ for the **composition** of proportions and proportion chains respectively their corresponding intervals and chords - in this chapter 5 we consider merely the total proportion of these compositions, and then the sum-symbol $\oplus$ acts on intervals and it assigns a new **interval** (and not the composition) to two ore more intervals - called their **"sum"**.

Despite this correspondence, there are plausible reasons for calling one 'sum' and the other one 'product'. We see the correspondences, for example, in both juxtapositions

$$[1,2] \oplus [1,2] = [1,4] \; \rightleftharpoons \; (1:2) \odot (1:2) \cong 1:4,$$
$$[8,9] \oplus [5,6] \oplus [9,10] = [2,3] \; \rightleftharpoons \; (8:9) \odot (5:6) \odot (9:10) \cong 2:3.$$

It should also be mentioned that the inversion of a proportion $A \cong a : b$ to the proportion $A^{inv} \cong b : a$ – that is, the inversion of the numerical ratios – corresponds to the downward addition ('**subjunction**' $\ominus$) of intervals. In formulas:

$$[c,d] \ominus [a,b] = [c,d] \oplus [b,a] \; \rightleftharpoons \; (c:d) \odot (b:a).$$

Finally: In connection with the identification of proportions with intervals, we also frequently have to deal with the concept of '**difference**', to whom we devote a separate mathematical definition for the sake of clarity:

Definition 5.2 (Sum and Difference of Musical Intervals)
If $I_1 = [a_1, b_1]$, $I_2 = [a_2, b_2]$ and $I_3 = [a_3, b_3]$ are three **musical intervals** then we define the sum and the difference of these intervals, - according to the following equations and equivalences:

$$I_1 \oplus I_2 = [a_1, b_1] \oplus [a_2, b_2] \cong [a_1 a_2, b_1 b_2] \equiv \text{the sum of the intervals } I_1 \text{ and } I_2,$$
$$I_1 \ominus I_2 = I_1 \oplus I_2^{\,inv} = [a_1, b_1] \oplus [b_2, a_2] \cong [a_1 b_2, b_1 a_2] \equiv \text{the difference of } I_2 \text{ to } I_1,$$

Corollary : The basic equations of Interval $-$ Arithmetic

$$I_1 \oplus I_2 = I_3 \Leftrightarrow I_2 = I_3 \ominus I_1 \Leftrightarrow I_1 = I_3 \ominus I_2.$$

And if we have the situation that $I_2 = I_3 \ominus I_1$, then I_2 is called the **complementary interval of** I_1 in I_3 or **difference of** I_1 **to** (or **in**) the **interval** I_3 – or briefly: 'the difference of I_3 and I_1' and vice versa.

This specification also includes the case where I_2 itself is a downward interval and that I_1 is 'greater' than I_3 in the descriptive sense.

In the language of **proportions**, this construction corresponds to the fusion model:

If $A \cong a_1 : a_2$ and $B \cong b_1 : b_2$ are given proportions and if $X \cong x_1 : x_2$ is a solution of the proportion equation. $X \odot A \cong B$, then the "interval" $X = (x_1, x_2)$ is the complementary interval of the "interval" $A = (a_1 \; a_2)$ in the "interval" $B = (b_1, b_2)$, in formula:

(continued)

Definition 5.2 (continued)

$$X \odot A \cong B \quad \Leftrightarrow \quad X \cong B \odot A^{\text{inv}} - \text{proportion model}$$
$$\rightleftarrows X \oplus A \cong B \quad \Leftrightarrow \quad X \cong B \ominus A - \text{musical interval model.}$$

In words: The musical interval $[x_1, x_2]$ is the difference between the interval $[a_1, a_2]$ and the interval $[b_1, b_2]$.

This interpretation is also valid the other way around.

Attention please: At this point we want again to emphasise that we have used the same symbol $\oplus$ for the **adjunction** in form of the **composition** of two or more proportion chains as well for the **adjunction** in form of the **sum** (or addition) of two (or more) proportions. To emphasise the difference again: In "composition", two proportion chains are put togheter by achieving, by means of suitable multiplication of both chains, that the two connecting magnitudes are equal. They can then be immediately arranged in series and put togheter. In addition to the outer members (magnitudes) the composition therefore also contains all intermediate members of both chains as well as the identical connecting magnitude. In contrast with the "sum" of two proportions (which can be also understood as 1-step proportion chains) we only get the outer (or "total") proportion of the composition of these two chains - the only intermediate member is missing, which means that the adjunction is again a proportion - and not a multiple membered proportion chain (with more than one step). Mathematically speaking, the sum-symbols $\oplus$, $\ominus$ acts in the set of all musical intervals and the operation fulfils all propoerties of an algebraic, commutatative group - which means, that we can calculate as in the case of integers. This concept of Adjunktion in form of Addition togheter with the invers difference ($\oplus$, $\ominus$) cannot be transferred to compositions of Proportion chains. One of several reasons for this is the fact, that the number of steps increases with each composition - as we now. For example: The composition $(1 : 2) \oplus (2 : 1)$ of the 1-step-chains $(1 : 2)$ and $(2 : 1)$ is the 2-stepped chain $(1 : 2 : 1)$, whereas the sum of these intervals is an interval - namely the unison $(1 : 1)$.

The Monochord

The almost sole purpose of experimenting with a single fixed string ('monochord') is to describe the dependence of pitches on the choice of an intermediate point. If, by means of such an intermediate point, we divide the given string (of length L_1) into two complementary parts (lengths L_x and L_{1-x}), we are asked how the tone above the sub-string L_x relates to the fundamental of the empty string or to the tone above the complementary string. Fig. 5.1 shows us this model.

This dependence is now describable by a plausible and simple – as well as modern-physical explainable – law of experience:

Fig. 5.1 The abstract
monochord model

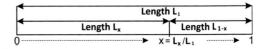

Proposition 5.1 (Monochord Laws)
If a tensioned empty string L_1 has the fundamental frequency f_1, and if we have the
division of the string into the two parts L_x and L_{1-x} according to the sketch, then the
following formula applies to the frequency f_x of the sub-string of length L_x

$$L_x * f_x = L_1 * f_1 = \text{const.} \equiv \text{Monochord frequency formula}$$

Conclusion: The laws of proportion apply

$$f_1 : f_x \cong L_x : L_1 \cong x : 1 \text{ and } f_{1-x} : f_x \cong L_x : L_{1-x} \cong x : (1-x).$$

They describe the relationships between pitches and string lengths; these are
inversely proportional to each other. For intervals follows the relation

$$[f_{1-x}, f_x] = [f_1, f_x] \ominus [f_1, f_{1-x}] \equiv \text{Monochord intervall formula}$$

Let us leave the monochord frequency formula as an empirical law, otherwise it would
be easy to derive from physical rules such as Mersenne's frequency law. Since the formula
of the constancy of the product of pitch and string length applies with equal justification to
the rest of the string L_{1-x} with its pitch f_{1-x}, we obtain everything that has been said from
the equation

$$L_x * f_x = L_{1-x} * f_{1-x}.$$

The interval balance can also be very quickly derived from the proportion identity

$$(f_1 : f_x) \odot (f_{1-x} : f_1) \odot (f_x : f_{1-x}) \cong (1 : 1)$$

whose translation into the language of intervals leads exactly to this balance, if one
converts this proportion equation according to the proportion $(f_{1-x} : f_x)$, which is the
inverse of the proportion. $(f_x : f_{1-x})$

Table 5.1 Harmonic divisions of the monochord

$[f_1 : f_x]$	ct value	Interval name	$[f_1 : f_{1-x}]$	ct value	Interval name
$1:1$	0	Unison (empty string)	–	–	–
$1:2$	1200	Octave	$1:2$	1200	Octave
$1:3$	1902	Fifth over octave	$2:3$	702	Fifth
$1:4$	2400	Double octave	$3:4$	498	Fourth
$1:5$	2786	Third over two octaves	$4:5$	386	Major third
$1:7$	3369	Nature seventh over two octaves	$6:7$	267	Nature Septime $\ominus$ Fifth
$1:8$	3600	Triple octave	$7:8$	231	Ekmelian whole tone

▶   It should be clear that this relation which seems almost trivial, is nevertheless a very essential one: the monochord combines **non-measurable** musical proportions with **measurable** geometric proportions.

If the string is halved, then both parts have double the frequency: the octave is realized. Certainly, in ancient times, the divisions of the monochord string were primarily thought of as corresponding to an integer division; thus the harmonic sequence (1/2, 1/3...) was chosen for the parameter x. Table 5.1 describes for such $x = 1/n$ some distinctive intervals that were certainly obtained in this way.

To get a little routine in this kind of intervallic calculation, one can both determine the size of the parts among themselves from this and confirm it by a separate calculation. The following example just illustrates what has just been said.

Example 5.1 Intervals of the Pythagorean Tonal System: The Primes 2 and 3

1. For $n = 3$ – i.e. for the division parameter $x = 1/3$ – we have the calculation

$$[f_{1-x}, f_x] = [f_1, f_x] \ominus [f_1, f_{1-x}] = [f_1, f_x] \oplus [f_{1-x}, f_x] \Longleftrightarrow$$
$$[1,2] = [1,3] \ominus [2,3] = [1,3] \oplus [3,2]$$
$$\Longleftrightarrow \text{octave} = \text{duodecime (twelfth) minus fifth.}$$

2. For $n = 9$ we find the pythagorean whole tone $8:9$ as an interval $[f_1, f_{1-x}]$ above the larger string section; and the representation of the interval

$$[1,9] = [1,8] \oplus [8,9]$$

explains its position as a whole tone over three octaves and at the same time confirms the monochordal interval formula.   ◀

The handling of the monochord – a 'simple, inconspicuous instrument' – can, with refined handling, drive and accompany the play instinct of explorative discovery very deeply into the scenery of music-theoretical problems; consider the succeeding example to underline the statement

---

Example 5.2 Iterated Applications of Monochordal Play

1. The **small - ore -minor diësis** is the 'difference of three perfect major thirds from the octave'; it has the proportion 125 : 128. (This will be discussed in Sect. 5.2).

   Can you make this interval audible on the monochord?

   At the moment we imagine the monochord provided with a second fundamental string of the same pitch – thus we have a '**duochord**'. We know that we come to the small diësis via the exponentiation of the proportion 4 : 5:

   $$(4 : 5) \odot (4 : 5) \odot (4 : 5) \cong 4^3 : 5^3 = 64 : 125.$$

   We start with the $n = 5$ -division of L_1; then the larger side $L_{4/5}$ has the perfect major third above the base string L_1.

   We fix this sub-point and declare this sub-string $L_{4/5}$ to be the 'new' fundamental string. We also divide this string in the $n = 5$ division; the new, larger piece again possesses the third above the new fundamental – in total we have then gained two layered thirds via L_1. If we repeat this process with the achieved division, we have obtained the perfect major third over the string L_1

   How long is the sub-string after the third iteration? It obviously is

   $$L_{4/5 * 4/5 * 4/5} = L_{64/125}.$$

   Now, on the other hand, we form the octave to the fundamental frequency on the reference string by halving, this is then the sub-string $L_{1/2} = L_{64/128}$, and we have the result: the difference of both tones forms the proportion

   $$(64 : 128) \odot (125 : 64) \cong (125 : 128) - \text{small Diësis}.$$

2. One could proceed in a similarly ingenious way to make even the tiny **syntonic comma** measurable and concretely constructible:

   First of all – this will also be discussed in the following Sect. 5.2 – this comma is the difference between four perfect fifths 2 : 3 and the pure major third 4 : 5 over two octaves and therefore has the proportion 80 : 81.

The fifth to the fundamental frequency is obtained as a tone over the 2/3 string $L_{2/3}$. With a fourfold iterated procedure – as in the previous example – we have obtained the sub-string $L_{16/81}$. Afterwards we only have to set up the five division on the reference fundamental string and then exactly the perfect major third sounds over the smaller $L_{1/5}$ string according to the monochord formula over two octaves. The difference between the two steps provides the syntonic comma

$$(16 : 81) \odot (5 : 1) \cong (80 : 81) = \text{syntonic comma.}$$

◀

The last example clearly shows that the playful handling of the monochord can be taken into the theory of tonal systems quite reasonably – especially due to a practice that monitors the theory. This is also the case for other methods of setting up tone or interval systems that are regarded as 'fundamental'. We would like to introduce some of these in the following:

Methods of Generating Interval Systems

Of the methods of finding intervals by which musical constructs such as scales and chords are erected, the systematic ones are particularly interesting; they possess the higher degree of classification in more general **principles.** We enumerate:

1. **The principle of iteration (or progression)**
   Intervals – or their proportions – are formed by the layering (addition, adjunction. . .) of two or more given generating **basic intervals** or by the product (fusion) of given proportions and their inverses.
   (a) In the pythagorean case, this is the conglomerate of all chain formations of fifths $2 : 3$ and octaves $1 : 2$ – or alternatively of fifths and fourths $3 : 4$ – the octave complement of the fifth.
   (b) In the case of just (diatonic) temperament, the generators are octave $1 : 2$, fifth $2 : 3$ (or alternatively fourth $3 : 4$) and third $4 : 5$.
   Thus, the two families of proportions

   $$\mathbb{P}_{\text{pyth}} = \{A | A \cong (\odot \text{ fifth})^m \odot (\odot \text{ octave})^n \text{ with } n, m \in \mathbb{Z}\}$$
   $$\mathbb{P}_{\text{diat}} = \{A | A \cong (\odot \text{ third})^k \odot (\odot \text{ fifth})^m \odot (\odot \text{ octave})^n \text{ with } k, n, m \in \mathbb{Z}\}$$

   are described as integer iterations of their two and three generating proportions respectively (see Sect. 1.4 for the symbolism).

2. **The principle of equidistant decomposition**

One divides the string (or a sound tube) into n equal (= equally long) parts and, by means of the monochord formula, obtains n step intervals, which result as the difference of adjacent sections. For the classical case of an eighth note of the fundamental string L_1 achieved by continuous halving, we then have the sequence of differences of the section lengths, which can obviously be written as an **arithmetic chain of the string division**

$$1 : 2 : 3 : 4 : 5 : 6 : 7 : 8$$

According to the monochord formula, the reciprocal chain for this corresponds exactly to the reverse order of the tone steps and according to our theory (Theorem 3.5), this reciprocal chain

$$(7 : 8) \oplus (6 : 7) \oplus (5 : 6) \oplus (4 : 5) \oplus (3 : 4) \oplus (2 : 3) \oplus (1 : 2)$$

is a **harmonic** chain with a sequence of intervals, the first two steps of which are **ecmelic** – but together produce a fourth $6 : 8$; the further steps are traditionally justly diatonic, so that at tonic C we have the tone series

$$c_0 - d_0^* - f_0 - as_0 - c_1 - f_1 - c_2 - c_3$$

The interval $\left[c_0, d_0^*\right]$ is with ≈ 231 ct clearly larger than a usual whole tone interval with 200 ct. The task of writing the harmonic chain as a connected chain of numbers is quite sportive due to the numerical ratios – one obtains:

$$(1 : 2 : 3 : 4 : 5 : 6 : 7 : 8)^{\text{rez}} \cong 105 : 120 : 140 : 168 : 210 : 280 : 420 : 840.$$

Here, too, one can convince oneself – if the theory is not being trusted – that every consecutive 3-step chain is harmonic, which is why the whole chain is a harmonic chain of proportions.

3. **The principle of proportional division**

This is understood to mean dividing the sounding string not 'equidistantly' but 'proportionally' according to the law of harmonic series: For a heptatonic scale, this is how we would proceed in our monochord model:

We choose for the division parameter x the value series

$$0 - \frac{1}{8} - \frac{1}{7} - \dots - \frac{1}{2},$$

Subsequently a sequence of proportions of the residual lengths L_{1-x} to L_1 results in the form

$$(7:8) - (6:7) - (5:6) - (4:5) - (3:4) - (2:3) - (1:2).$$

This time we do not attach them to each other (to form a harmonic chain), but we follow the sequence of tones by having each interval act from the tonic C: this way, together with the tonic $(x = 0)$, we obtain the ascending sequence of tones of an octave scale, which we also quote together with the (rounded) cent numbers of the intervals applied from the tonic:

$$c - d^* - \mathrm{dis}^* - es - e - f - g - c \quad | \quad 231 - 266 - 315 - 386 - 498 - 702 - 1200.$$

The step proportions of this scale are calculated as the differences of adjacent intervals – thus we obtain – from the 2nd step onwards – the step series ascending from tonic C as a remarkable chain of proportions of the step series

$$\left((n^2 - 1) : n^2 \right), n = 7, 6, \ldots, 2,$$

which then concretely leads to the composition

$$(7:8) \oplus (48:49) \oplus (35:36) \oplus (24:25) \oplus (15:16) \oplus (8:9) \oplus (3:4).$$

If the number of divisions were increased to any number, the result would indeed be a sequence of steps striving towards the unison $1:1$ if the scale were followed downwards – i.e. having its reciprocal

$$(3:4), (8:9), (15:16), \ldots, \left((n^2 - 1) : n^2 \right), \ldots \to (1:1) \text{ for } n \to \infty$$

in view. By the way, the intervals of the tone sequence over the sub-sections

$$L_{1/8}, L_{1/7}, \ldots, L_{1/2}, L_1$$

to the fundamental tone of the empty string would again form an arithmetic chain of proportions – indeed, the relationship is the equivalence of

$$A = \frac{1}{8} : \frac{1}{7} : \frac{1}{6} : \frac{1}{5} : \frac{1}{4} : \frac{1}{3} : \frac{1}{2} : \frac{1}{1} \Leftrightarrow A^{\mathrm{rez}} \cong 1 : 2 : 3 : 4 : 5 : 6 : 7 : 8.$$

Historical   By the way, this system was realized in the **chinese zither;** this actually has this strange sequence of steps. Another observation is that some intervals of the greek-antique tetrachordic had quite exotic steps. So here the intervals $(48:49)$ as well as

(35 : 36) are not unfamiliar. It may well be that the monochord construction of harmonic string sections in the past led to these proportions. It would be conceivable.

4. **The principle of consonant division**

The notion of **consonance** is strongly related to the notion of so-called 'simple-superparticular' proportions ($n : n + 1$) or the corresponding intervals [$n, n + 1$]. One occasionally encounters this definition (cf. [16]):

> An interval [n, m] with integer magnitudes n, m is **consonantly divided** if it is composed as a sum of simple-superparticular intervals.

However, it should be noted that by a trick you can divide any interval [$n, m = n + k$] simply-superparticularly; you simply decompose it in this form:

$$[n, m = n + k] = [n, n + 1] \oplus [n + 1, n + 2] \oplus \ldots \oplus [m - 1, m].$$

For the proportions this means the simple-superparticular decomposition into the products

$$(n : m) = (n : n + 1) \odot (n + 1 : n + 2) \odot \ldots \odot (m - 1 : m).$$

Now it is interesting that also simple-superparticular proportions themselves can be written again as a chain of proportions consisting of **any number of simple-superparticular steps** – this is conceivably simple and it is enough for us to show this once for a decomposition into two proportions: Namely, the balance holds

$$(n : n + 1) \cong (2n : 2n + 2) \cong (2n : 2n + 1) \odot (2n + 1 : 2n + 2).$$

Subsequently the corresponding proportion chain of composition $2n : (2n + 1) : (2n + 2)$ is the simple-superparticular decomposition of the proportion ($n : n + 1$). With the new two sub-proportions, one can repeat this process and in this way achieve a **consonant division of** a simple-superparticular proportion into any number of consonant parts.

Historical In the antiquity up to the Middle Ages there was an unmanageable abundance of such divisions; Boethius (Anicius Manlius Severinus Boethius ($\approx 480 - 525$)) divided the Tonos (8 : 9) into the parts (16 : 17) and (17 : 18). As he recognized that both parts were different (we would say: not similar), he hereby justified the 'indivisibility' of the tonos (into two equal halves) and thereby supported – unfortunately, however, with a faulty argument – the theses of the pythagorean doctrine, namely that the 'tonos' was indivisible.

5. **The principle of division by means of the (classical, babylonian) medieties**

As the example of the arithmetic proportion chain $2n : (2n + 1) : (2n + 2)$ from the previous consideration shows directly, a simple insertion of the arithmetic mean has been done in this case. Without doubt, there are many more possibilities to extend this model to all possible averages – either by introducing inner medieties (to the magnitudes n and m) or by calculating diverse outer proportionalities – according to the models of our Sect. 3.4. These constructions can be encountered by seeking after.

5.2   Musical Tone Systems: The Proportion Equation as a Path to Harmony

In this section we turn to concrete and historically relevant tonal systems, especially those created by building blocks and their iterative adjunction and subjunction. At the same time, it serves to describe intervals by their proportions. We now mainly obtain these from the calculus of the fusion equations of Sect. 1.4. In three larger blocks of examples, we present the most important intervals

- of the pythagorean fifth system $\mathbb{P}_{pyth}$,
- of the perfect diatonic fifth-third system $\mathbb{P}_{diat}$,
- of an ecmelian interval system, which requires the prime numbers 7 and 11,

The former two find rich occurrences, especially in the familiar theory of scales and in the theory of musical commas, while the more remote 'ecmelic' intervals can cause astonishment: We describe

- *both an **illustrious connection to the circular number π**,*
- *as well as relations to **ancient tetrachordics.***

The latter, as is well known, was exceedingly rich in unusual structures of proportions.

In the first example block, we develop the architecture of the **pythagorean system,** that is known to be built exclusively of fifths $(2 : 3)$ and octaves, if we use the language of intervals. In the language of proportions, this then means that all the proportions to be obtained $(n : m)$ can only be constructed from the

- **arithmetic building blocks:** unison $(1 : 1)$, octave $(1 : 2)$ and octavated fifth (duodecime) $(1 : 3)$ and their inversions, the
- **harmonic components:** unison $(1 : 1)$, downward octave $(2 : 1)$ and downward duodecime (twelfth) $(3 : 1)$

by means of addition (product, fusion). The usual fifth $(2 : 3)$ itself is a construction,

Fifth $(2 : 3) = $ Duodecime $(1 : 3)$ minus Octave $(1 : 2) = (1 : 3) \odot (2 : 1)$.

Nevertheless, it is considered to be the basic building block of almost all justly-tuned musical systems.

Example 5.3 Intervals of the Pythagorean Tonal System: The Primes 2 and 3

Generating intervals of all other intervals from $\mathbb{P}_{pyth}$ are the perfect fifth $2 : 3$ and the octave $1 : 2$. As usual, we denote in some equations the fith by the symbol Q (coming from latin: quintus- the fifth) and the octave by the symbol O.

1. **Tonos (major whole tone):** As we have already seen in Sect. 1.4, Example 1.2, the proportions of the pythagorean whole tone (tonos T) as a layering of two fifths (minus an octave) and the major pythagorean third **(ditonos)** as a layering of two whole tones T are these:

$$\text{Tonos} = 2 \text{ fifth} \ominus \text{ octave} = Q \oplus Q \ominus O = (2 : 3) \odot (2 : 3) \odot (2 : 1) \cong 8 : 9$$
$$\text{Ditonos} = 2T = T \oplus T = (8 : 9) \odot (8 : 9) \cong 64 : 81.$$

   The tonos has the rounded logarithmic measure of ≈ 204 ct.

2. **Limma:** What is the proportion of the interval that results as the difference of a major pythagorean third to the fourth?

   The perfect fourth has the proportion $3 : 4$ because it forms together with the fifth $2 : 3$ an octave $1 : 2$ – which is indeed confirmed by

$$(2 : 3) \odot (3 : 4) \cong 6 : 12 \cong 1 : 2$$

   Accordingly, for the interval X we are looking for, we have the equation

$$X \odot (64 : 81) \cong 3 : 4 \Leftrightarrow X \cong (3 : 4) \odot (81 : 64) \cong 243 : 256 = 3^5 : 2^8.$$

   This interval to the proportion $243 : 256$ is called **pythagorean Limma (L).** And we have the interval balance

$$L = (O \ominus Q) \ominus 2T.$$

   Since a tonos has the proportion $8 : 9 \cong 243 : 273\frac{3}{8}$, the limma is only almost half the size of the tonos; this can be seen more easily in the cent measure: the limma has the rounded logarithmic measure of ≈ 90 ct.

3. **Apotome:** What is the size of the limma's partner in the tonos?
   The corresponding equation is:

$$X \odot \left(3^5 : 2^8\right) \cong 8 : 9$$
$$\Leftrightarrow X \cong \left(2^3 : 3^2\right) \odot \left(2^8 : 3^5\right) \cong \left(2^{11} : 3^7\right) = 2048 : 2187.$$

This interval is called **pythagorean Apotome (A)**; we thus have the decomposition of the whole tone tonos into two (differently sized) semitonia,

$$\boldsymbol{T = L \oplus A} \rightleftarrows (8 : 9) \cong (243 : 256) \odot (2048 : 2187).$$

The apotome has the rounded logarithmic measure of $\approx 104\,\mathrm{ct}$. Hence, the apotome is the larger and the limma is the smaller half of the tonic.

4. **Pythagorean comma:** Finally, what is the difference between these two semitonia A and L? Again, we obtain the result from the necessary equation of proportions:

$$X \odot \left(3^5 : 2^8\right) \cong \left(2^{11} : 3^7\right) \Leftrightarrow X \cong \left(2^{11} : 3^7\right) \odot \left(2^8 : 3^5\right) = \left(2^{19} : 3^{12}\right)$$
$$\Leftrightarrow X \cong 524.288 : 531.441 \rightleftarrows X = A \ominus L,$$

and this is the **Pythagorean comma** - respectively **"the comma of Pythagoras"** -, whom we had already encountered in the example block of Sect. 1.4. Its logarithmic size is $\approx 23, 5\,\mathrm{ct}$.

5. **The Pythagorean heptatonic scale:** If we layer five upward fifths and separately one downward fifth from a starting note (tonic) and bring the achieved notes into the octave space above the tonic by means of suitable octave scaling, we obtain a 7-step (heptatonic) octave scale, which necessarily has the following ascending interval steps of whole tones T and semitones L of the familiar pattern of a major scale

$$T - T - L - T - T - T - L - \text{in short}: 1 - 1 - \frac{1}{2} - 1 - 1 - 1 - \frac{1}{2}$$

The Apotome does not appear here, it is not a member of the heptatonic scale – but it is a step interval in the 12-step chromatic scale. The proportion chain constructed according to this heptatonic structure as a multiple adjunction of the step proportions from tonos $8 : 9$ and limma $243 : 256$ is then as follows

$$384 : 432 : 486 : 512 : 576 : 648 : 729 : 768.$$

This is the proportion chain of the **pythagorean heptatonic** of smallest possible integer magnitudes. We see – quasi as confirmation of the calculation – the octave balance of the outer magnitudes $384 : 768 \cong 1 : 2$. ◀

In order to give the reader some help in creating this long chain, we will carry out a few steps in this regard: the composition of the seven proportions

$$(8 : 9) \oplus (8 : 9) \oplus (243 : 256) \oplus (8 : 9) \oplus (8 : 9) \oplus (8 : 9) \oplus (243 : 256)$$

should thus be calculated in exactly this order as a 7-step proportion chain. We see, on the one hand, the symmetrical construction of two equal tetrachords connected by a whole tone step

$$[(8 : 9) \oplus (8 : 9) \oplus (243 : 256)] \oplus (8 : 9) \oplus [(8 : 9) \oplus (8 : 9) \oplus (243 : 256)];$$

and on the other hand, we will work with the least common multiple (lcm) at the junctions of two subchains to be attached, taking into account the prime factor structure of the magnitudes. We start with the chain of the tetrachord:

$$(8 : 9) \oplus (8 : 9) \cong (64 : 72) \oplus (72 : 81) \cong 64 : 72 : 81 = 2^6 : 2^3 3^2 : 3^4$$

Then we add the limma $(243 : 256 = 3^5 : 2^8)$ – here apparently the tripling of the front chain is sufficient – and then for the tetrachord we get the balance

$$\left(2^6 3^1 : 2^3 3^3 : 3^5\right) \oplus \left(3^5 : 2^8\right) \cong 2^6 3^1 : 2^3 3^3 : 3^5 : 2^8.$$

For the following addition of the whole tone $2^3 : 3^2$ to this tetrachord, the whole tone is extended with 2^5 to the similar proportion $2^8 : 2^5 3^2$, hence we get the **pentachord**

$$\left(2^6 3^1 : 2^3 3^3 : 3^5 : 2^8\right) \oplus 2^8 : 2^5 3^2 \cong 2^6 3^1 : 2^3 3^3 : 3^5 : 2^8 : 2^5 3^2.$$

Now this pentachord is extended by 2^1 and the tetrachord by 3^1. Therefore the connecting magnitudes are identical again and we obtain with

$$\left(2^7 3^1 : 2^4 3^3 : 2^1 3^5 : 2^9 : 2^6 3^2\right) \oplus \left(2^6 3^1 : 2^3 3^4 : 3^6 : 2^8 3^1\right)$$
$$\cong \left(2^7 3^1 : 2^4 3^3 : 2^1 3^5 : 2^9 : 2^6 3^2 : 2^3 3^4 : 3^6 : 2^8 3^1\right)$$

the required series $384 : 432 : 486 : 512 : 576 : 648 : 729 : 768$.

Two interesting **remarks** may be mentioned at this point:

1. The familiar pattern of the major scale

$$1 - 1 - \frac{1}{2} - 1 - 1 - 1 - \frac{1}{2}$$

is not only the familiar **pattern of the whole tone and semitone sequence** of our usual scale, but it is the pattern of any heptatonic (major) scale produced by iterations of a fixed and arbitrary fifth. If, for example, one were to take the mean-tone fifth (of irrational proportion $1 : \sqrt[4]{5} \approx 1,4953.. \approx 696,58\,\text{ct}$), the mean-tone scale of **Michael Praetorius** and **Arnold Schlick** would be created, which would have the same sequence of whole and half tones (but of different sizes).

2. The **question** arises: Is it a coincidence that the semitone difference represents the same interval of the pythagorean comma as the difference of six whole-tone steps to the octave?

   The **answer** is: No, both are the difference of 12 perfect fifths to 7 octaves!

   Why? If we use for all intervals (T, L, A) their representations as stratifications of fifths and octaves, we get the same balance every time. (Please try it out!) A detailed discussion of this branch of music theory and its mathematical descriptions can be found in [16].

The next set of examples deals with the just-tuned **diatonic interval family** $\mathbb{P}_{\text{diat}}$; these are (according to the majority understanding) all intervals whose frequency measures can be expressed by the three prime numbers 2, 3, and 5. Musically and equivalently this means that to the two generating intervals octave $1 : 2$ and fifth $2 : 3$ of the pythagorean system $\mathbb{P}_{\text{pyth}}$ the perfect third $4 : 5$ is added. In the language of the theory of proportions, this interval – or rather this tonal system – is built up from the

- **arithmetic** proportion blocks prime $(1 : 1)$, octave $(1 : 2)$, duodecime $(1 : 3)$ and double octave-third $(1 : 5)$ together with their inversions,
- **harmonic building blocks** $(1 : 1)$, $(2 : 1)$, $(3 : 1)$, and $(5 : 1)$, which correspond to the downward added intervals of the arithmetic sequence.

▶   **Modern Aspect** *With tacit suppression of occasional octavations – which (as above) have the sole purpose of transporting tones obtained by interval stratifications, if necessary (and for example), into the octave space above the tonic as the starting tone – we thus move in the whole-numbered gridded 2-dimensional **eulerian** grid **of all** third $(4 : 5)$ – and fifth $(2 : 3)$ – iterations. In this grid, one can mathematically methodically reach – as it were in the form of an integer **vector calculation** – all intervals/tones of the just diatonic temperaments, of which there are indeed many.*

There are some outstanding intervals in this system by which the structure of whole, half, and quarter tones can be systematically described. Among these, we will pick out a few:

Example 5.4 Intervals of the just Diatonic – The Prime Numbers 2, 3, and 5

1. **Minor (diatonic) whole tone:** What is the proportion of the interval that results as the difference of the pythagorean whole tone (tonos) $8 : 9$ in the just third $4 : 5$?
   The defining proportion equation is given by the calculus

$$X \odot (8:9) \cong 4:5 \Leftrightarrow X \cong (4:5) \odot (9:8) \cong 36:40 \cong 3^2 : 2^1 5^1 = 9:10.$$

Hence, the just major third has the decomposition into **two** integers of **different sizes**

Third $(4:5) = $ major whole tone $(8:9) \oplus$ minor whole tone $(9:10)$;

the minor diatonic whole tone $9:10$ has the logarithmic measure $\approx 182, 4\,\mathrm{ct}$.

2. **Syntonic comma:** The difference between these two whole tones is called the syntonic (occasionally didymic) comma-respectively "the comma of Didymus"-; its proportion is quickly determined:

$$X \odot (9:10) \cong 8:9 \Leftrightarrow X \cong (8:9) \odot (10:9) \cong 80:81 = 2^4 5^1 : 3^4.$$

Thus we have the two mutually equivalent balancing formulas

Tonos $(8:9) = $ minor whole tone $(9:10) \oplus$ syntonic comma $(80:81)$,
Ditonos $(64:81) = $ major third $(64:80) \oplus$ syntonic comma $(80:81)$.

The syntonic comma is a little smaller than the pythagorean comma at $\approx 21, 5\,\mathrm{ct}$; the difference is a '**schism**' of $\approx 2\,\mathrm{ct}$, see (4).

3. **Diatonic semitone** S: What is the difference between the just third $4:5$ and the just fourth $3:4$?

On a diatonically just-tuned piano, for example, this would be the interval from e to f. The calculation yields

$$X \odot (4:5) \cong 3:4 \Leftrightarrow X \cong (3:4) \odot (5:4) = 3^1 5^1 : 2^4 \cong 15:16.$$

This semitone is called a just or perfect or pure (diatonic) semitone; it is larger than the pythagorean limma by the syntonic comma, and its logarithmic magnitude is $\approx 111, 7\,\mathrm{ct}$.

4. **Schism:** The difference of syntonic and pythagorean comma is

$$X \odot (80:81) \cong (2^{19} : 3^{12}) \Leftrightarrow X \cong (2^{19} : 3^{12}) \odot (3^4 : 2^4 5^1) \cong 2^{15} : 3^8 5^1.$$

Thus $X \cong 32.768 : 32.805$ – and this tiny interval is almost a unison, in the more handy logarithmic scale just $\approx 2\,\mathrm{ct}$. In historical musicology the schism was the smallest musical interval of the diatonic.

The abundance of elementary intervals only begins here: the diatonic semitone S has complementary partners in the minor as well as in the major whole tone – called **small** and **large chroma**; the difference of the small chroma in the major whole tone $(8:9)$ is in turn called **Eulerian semitone;** three just major thirds $(4:5)$ have a comma with the octave (the small **diësis**), just as four just minor thirds (5 : 6) form a comma difference with the octave (the large or **great diësis**). We calculate the proportions of all these intervals and hopefully shed some light on this interval chaos:

5. **Small** and **large chroma:** According to the preceding description, we have respectively the defining equations

$$X \odot (15 : 16) \cong 9 : 10 \Longleftrightarrow X \cong 9 * 16 : 15 * 10 = 24 : 25 \text{ (small chroma)},$$
$$X \odot (15 : 16) \cong 8 : 9 \Longleftrightarrow X \cong 8 * 16 : 9 * 15 = 128 : 135 \text{ (great chroma)}.$$

These, then, are the partners complementary to the diatonic semitone in the two whole tones; their logarithmic magnitudes are $\approx 70, 7\,$ct (for the small chroma) and $\approx 92, 2\,$ct (for the large chroma).

6. **Eulerian semitone:** Since the small chroma – as well as the large – have the character of a 'semitone' – since they are, after all, the complements of a 'semitone' in a 'whole tone' – another semitone difference can actually be formed in a whole tone: The difference of the small chroma in the major whole tone, and this results in

$$X \odot (24 : 25) \cong 8 : 9 \Longleftrightarrow X \cong 8 * 25 : 9 * 24 = 25 : 27;$$

this new semitone(step) is called Eulerian semitone; it has considerable $\approx 133, 2\,$ct.

7. **Small diësis** and **large diësis:** The balances here give the proportion values

$$X(\odot(4 : 5))^3 \cong 1 : 2 \Longleftrightarrow X \cong 5^3 : 2^7 = 125 : 128 \text{ (small diësis)},$$
$$X \odot (1 : 2) \cong (\odot(5 : 6))^4 \Longleftrightarrow X \cong 5^4 : 2^3 3^4 = 625 : 648 \text{ (large diësis)},$$

where $(\odot(a : b))^m$ is the *m-fold* proportion product (see Theorem 1.4).

The logarithmic measures are $\approx 41\,$ct (small diësis) and $\approx 62, 5\,$ct (large diësis); both differ by the syntonic comma.

8. **The just-tuned - or the perfect - or pure - (diatonic) heptatonic scale** is obtained from the Euler-grid of iterations with thirds (4 : 5) and fifths (2 : 3) (and suitable down-octavations); one – of many possible(!) octave scales – has the interval structure

Tonos – minor whole tone – diatonic semitone – Tonos
– Tonos – minor whole tone – diatonic semitone.

If we compose the corresponding proportions to a proportion chain, we obtain the proportion chain representation

$$24 : 27 : 30 : 32 : 36 : 40 : 45 : 48;$$

due to the smaller numerical values of its magnitudes, it is much more clearer than the proportion chain of the pythagorean heptatonic scale.   ◀

The following example leads us to a playground of ancient interval arithmetic, as one likes to encounter it in deepening older literatures and where surprises are usually as guaranteed as the certain loss of an overview of it.

When we mention the famous circle number π – with its approximation 3.14... – its inclusion in the realm of musical forms seems very remote. After all, it describes the ratio of the diameter to the periphery, the circular line. It has long been clear that π does not enter into a commensurable relationship n : m ; this number is irrational and, moreover, transcendent, which suggests a more complex structure.

Now, these circumstances were completely unknown in former times; the problems of the 'squaring of the circle' and many comparable things were only comprehensively recognized and solved in our more recent times. If we think of claims such as that of Pythagoras that everything is somehow commensurable, it is not surprising that attempts were also made to put a musical cloak on the number π.

Therefore we come to the '**ecmelian third**'. The starting point is the following: since forever, π has been calculated – indeed identified – with the fraction 22/7. A comparison shows:

$$\pi = 3,1415 \text{ and } 22/7 = 3,1428 - \text{ therefore } (22/7 : \pi) = 1,000432\ldots.$$

Proportionally, the error difference is hence just half a thousandth after the decimal point. This convenient ancient substitution is predestined for common practice – especially since its magnitudes (22 and 7) are very convenient (if we put the calculator aside).

If we replace π by this formidable approximation, we get an **ecmelic proportion** (7 : 22), which mutates into the proportion (7 : 11) when octaved down. It thus contains exactly the two primes 7 and 11 which follow the "diatonic" prime numbers 2, 3, and 5 of the senarium. From this proportion (7 : 11) one has then constructed the following tones or intervals:

Example 5.5 The Circular Number π and the Ecmelian Thirds: The Primes 7 and 11

Adding the next two prime numbers (7 and 11) to the prime numbers 2, 3, and 5 of the pythagorean diatonic system leads to very exotic intervals; some of which are these:

1. **Ecmelian whole tone:** The whole tone step 7 : 8 with 231, 2 ct is already considerably larger than the already very large tonos (8 : 9) with 203, 9 ct.
2. **Ecmelic sixth:** The proportion 7 : 11 belongs to an interval whose cent measure is 782, 5 ct. If we place it in an equal scale, this interval can be considered a minor sixth (with 800 ct).
3. **Ecmelian fourth:** the proportion 8 : 11 describes an interval with 551, 3 ct, whose position can (but does not have to) assume the role of a (quite large) fourth; its octave complement then has the proportion 11 : 16 and would represent the corresponding **Ecmelian fifth** (with 648, 7 ct).
4. **Ecmelian thirds:** Due to the fact that their proportions and measures differ considerably from those of the usual tone steps, several variants of ecmelian thirds can be given:
   (a) As an octave complement to the minor sixth 7 : 11 we have the proportion of the

   $$\text{Third}_{\text{ecmelic}}\ (11 : 14)\ \text{with the measure 417, 5 ct.}$$

   (b) As a layering (product) of two ecmelic whole tones 7 : 8 we obtain the

   $$\text{Third}_{\text{ecmelic}}\ (49 : 64)\ \text{with the measure 462, 4 ct}$$

   (c) The difference of the ecmelic whole tone 7 : 8 in the ecmelic quart 8 : 11 is also obtained by means of the proportion equation:

   $$X \odot (7 : 8) \cong 8 : 11 \Longleftrightarrow X \cong (8 : 11) \odot (8 : 7) \cong 64 : 77$$
   $$X = \text{minor Third}_{\text{ecmelic}}\ (64 : 77)\ \text{with the measure 320, 14 ct.}$$

   ◀

Surprising **coincidence:** this minor ecmelic third 64 : 77 coincides, astonishingly, almost exactly with one of the minor thirds found in the extended chromatic pythagorean scale – namely with the interval c – **dis** of the chromatic, pythagorean wolf-fifth scale generated by perfect fifths 2 : 3; in this, the interval $(c - \text{dis})$ has the proportion

$$(c - \mathrm{dis}) = 2^{14} : 3^9 = 16.384 : 19.683 \text{ with the measure } 317, 6 \text{ ct},$$

so that this minor ecmelic third is not completely remote from familiar tonal systems; nevertheless, its distance from the equal-tempered minor third (300 ct) is, again, considerable.

It should be noted that one surely suspects what further abundance of different whole tones, semitones and microtones could be formed if one were to list all possible differences of these intervals among themselves and together with the already known diatonic structures. In fact, some of these seemingly exotic proportions are ancient property. For example, the proportion series of **Ptolemy**'s **lydian diatonic tetrachord** is as follows

$$(7 : 8) - (9 : 10) - (20 : 21),$$

which corresponds to the proportion chain of the tetrachord 63 : 72 : 80 : 84, whose outer proportion 63 : 84 is indeed exactly a fourth 3 : 4. We will come back to this in the next Sect. 5.4.

5.3 Chordal Sounds in the Proportion Chains

In this section, we will present a few examples of musical realizations of strung-together proportions – i.e., chains of proportions; here, the inclusion of one or another medietary structure is also highly interesting.

Starting with the elementary but profound interrelations of arithmetic and harmonic proportion chains, which were originally expressed in the theorem on the harmonia perfecta maxima babylonica (Sect. 3.2) with the major-minor mirroring, and which – seen in the light of day – have accompanied us throughout the book. Afterwards we look at the major-minor symmetry in the complete **senarius** – the arithmetic building block of the third-fifth diatonic – and ask *how far the chords could be extended by external proportionals within the senarius.*

Finally, two more examples from playing with **sixth and seventh chords** and their proportion chains follow.

| Example 5.6 Major and Minor Triads of the Diatonic and Their Proportion Chains |

1. The **major triad of** the just diatonic temperament $\mathbb{P}_{\mathrm{diat}}$ – for example the tone series $c - e - g$ – is described by the 3-membered proportion chain 4 : 5 : 6; it has the balance of a fifth (2:3) and the step structure

major third $(4 : 5) \oplus$ minor third $(5 : 6)$;

the proportion chain is arithmetic.

2. The minor triad of the just diatonic temperament $\mathbb{P}_{\text{diat}}$ – for example the tone series $c - es - g$ – has the following structure, which is reciprocal to the major triad,

minor third $(5 : 6) \oplus$ major third $(4 : 5)$;

the 3-membered proportion chain results very quickly according to the adjunction procedure presented in Theorem 2.4 and is $10 : 12 : 15$, since

$$5 : 6 \cong 10 : 12 \quad \text{and} \quad 4 : 5 \cong 12 : 15.$$

The chain $10 : 12 : 15$ is a harmonic chain of proportions.

3. The two proportion chains of diatonic major and minor

$$4 : 5 : 6 \ \ (\text{major}) \quad \text{and} \quad 10 : 12 : 15 \ \ (\text{minor})$$

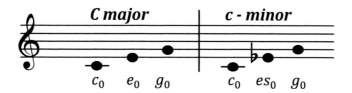

are – by construction – reciprocal to each other, which can also be shown by the numerical values – as an exercise.

4. The arrangement of the babylonian means can also be compared very well if we use the same outer proportion of the just fifth $(2 : 3)$ for both chains, which have the same outer proportion; after a short consideration, this is $20 : 30$; Hence follows

$$20 : 25 : 30 \ \ (\text{major triad}) \quad \text{and} \quad 20 : 24 : 30 \ \ (\text{minor triad}),$$

and 25 is the arithmetic and 24 the harmonic mean of the magnitudes 20 and 30. The connecting interval $[24, 25]$ is also quickly identified: it is the

small chroma $24 : 25$,

the difference between the minor diatonic whole tone and the diatonic semitone; we already encountered it in Example 5.4 – see also the table in the appendix.◀

In the next set of examples, we expand these major and minor chords – to the extent that this is possible both within the framework of the third-fifth grid $\mathbb{P}_{\mathrm{diat}}$ of Euler and within the framework of characterization as arithmetic and harmonic proportion chains.

The **action space of Senarius** is the totality of all proportion chains, which can be derived by adjunctions and products (fusions) from the Perissos-Artios-proportions

$$(1:1),(1:2),\ldots,(1:6) \equiv \text{Perissos} - \text{Proportions}$$
$$(6:1),(5:1),\ldots,(1:1) \equiv \text{Artios} - \text{Proportions}$$

Octave, fifth and just major third are therefore the corresponding intervals that can be formed from this – consequently, the action space of Senarius includes precisely all those proportions that can be obtained as intervals from these three building blocks via adjunction and subjunction.

Example 5.7 Arithmetic and Harmonic Proportion Chains of Senarius

---

1. The 3-membered chain $4:5:6$ can, within the senarius as an **arithmetic** chain, obviously only be continued maximally up to the 6-membered chain

$$1:2:3:4:5:6$$

With respect to the initial chain $4:5:6$, the magnitudes 3, 2 and 1 in this series are the successive outer (front) higher proportionals. This proportion chain describes a 6-note major chord in the step series

Octave $(1:2) \oplus$ Fifth $(2:3) \oplus$ Fourth $(3:4) \oplus$ major Third $(4:5)$
$\oplus$ minor Third $(5:6)$,

and it would be, for example, realized through the spread chord

$$c_0 - c_1 - g_1 - c_2 - e_2 - g_2$$

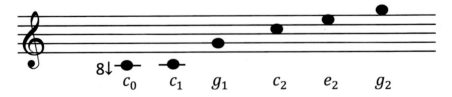

that can be found a million times over in the piano keyboards.

2. Consequently, the harmonic minor triadic chain $10 : 12 : 15$, which is reciprocal to the major proportion chain, can only be continued upwards as a harmonic chain by means of further outer proportions – exactly three times. This is described by Theorem 4.1. The reciprocal chain to the arithmetic chain $1 : 2 : 3 : 4 : 5 : 6$ is the **harmonic** proportion chain

$$10 : 12 : 15 : 20 : 30 : 60.$$

For example, it would be realized on the white keys of a piano keyboard (in just temperament) by the chord

$$e_0 - g_0 - h_0 - e_1 - h_1 - h_2.$$

We can calculate this series of numbers either by, on the one hand, inverting the arithmetic chain and then aiming at integerity, or, on the other hand, by determining the values by appending higher harmonic proportionals to the initial chain $10 : 12 : 15$; for then $x = 20$ is the appropriate value, so that 15 is the harmonic mean of $12 : 15 : x$. Or else we **argue musically:**

Musical construction: We join the intervals of the arithmetic chain in reverse: On top of the already existing minor chain

$$\text{minor third } (5 : 6) \oplus \text{major third } (4 : 5) \rightleftarrows 10 : 12 : 15$$

follows the fourth $3 : 4$, which – added to 15 – shows the proportion $15 : 20$. This is followed by a fifth $(20 : 30)$, and then an octave $(30 : 60)$; in this way the proportion picture is created quite simply – **by the music itself.**   ◄

We note that the non-continuability of the minor chain as a harmonic chain beyond the magnitude 60 is absolutely impossible even without an argumentation by means of its reciprocals, since the last interval is an octave. In a harmonic proportion chain $a : x : b$, however, the proportion $a : x$ cannot be similar to $1 : 2$ (or greater); otherwise, a contradiction would very quickly arise, which we have already recognized many times; nevertheless, we will show it again 'proportionally' at this point:

First, we have the transformation

$$a : x \cong 1 : 2 \Leftrightarrow x : 1 \cong 2a : 1$$
$$\Rightarrow (x-a) : (b-x) \cong (2a-a) : (b-x) \cong a : (b-x).$$

For this calculation we have used the exchange rule (Theorem 1.3) and the multiplication rule. Now, if the magnitude x were the harmonic mean of a and b, thus the proportion

$$(x-a) : (b-x) \cong a : b,$$

would be fulfilled and according to the foregoing also

$$a : (b-x) \cong a : b$$

and therefore with the reduction rule also

$$(b-x) : 1 \cong b : 1$$

which would be a contradiction. We had already seen this limiting condition on the harmonic mean in Sect. 3.3 when introducing all these medieties.

We describe – for practicing application – two simple examples from the harmony of chords:

Example 5.8 Major and Minor: 4-Membered Proportion Chains and Their Reciprocals

Let there be three musical intervals given by proportions

$$a : b \cong 4 : 5, \; b : c \cong 3 : 4 \; \text{and} \; c : d \cong 2 : 3$$

Afterwards we form the proportion chain $a : b : c : d$, which corresponds to a four-note harmony. Therefore we define **the task:**

(a) Describe the chord musically.
(b) Calculate the matching number proportion chain.
(c) Construct the reciprocal chain and describe a chord that corresponds to this reciprocal.

The solution (to (a) and (b)): If the note c is the tonic, the interval $4 : 5$ of the major third begins the chord, above it comes a fourth with $3 : 4$ and above it follows a fifth with $2 : 3$. The result is the spread *A **minor** chord* (third position).

$$c - e - a - e' - \text{shortly} : \text{major third} \oplus \text{fourth} \oplus \text{fifth}.$$

We use the appending procedure to determine the numerical proportions: To append the proportion $3 : 4$ to the proportion $4 : 5$, we increase the former by a factor of 3 and the proportion $3 : 4$ by a factor of 5. With $12 : 15$ and $15 : 20$ we obtain two proportions with a common connecting element, hence the chain

$$12 : 15 : 20.$$

Now, to connect the proportion $2 : 3$ to this subchain, we just need to extend it to the similar proportion $20 : 30$. We get the proportion chain

$$a : b : c : d \cong 12 : 15 : 20 : 30$$

as a numerical description of the chord $c - e - a - e'$ in just tuning. It is a harmonic chain of proportions, as we can easily verify: 15 is the harmonic mean of 12 and 20, and 20 is the harmonic mean of 15 and 30.

To part c): We obtain the reciprocal number-proportion chain with the procedure of a denominator-free representation

$$(12 * 15 * 20) : (12 * 15 * 30) : (12 * 20 * 30) : (15 * 20 * 30).$$

Although we do not calculate this chain accordingly – we see the reduction factors $(10, 5, 3, 3, 2, 2)$ and get the similar, very concise chain

$$2 : 3 : 4 : 5,$$

that consists of the layering (adjunction) of fifth, fourth and third (in this order), in complete agreement with the initial proportion chain, in which the arrangement was reversed. Finally we have the spread **C major chord**

$$c - g - c' - e' \text{ shortly : fifth} \oplus \text{fourth} \oplus \text{major third,}$$

– formed on the tonic **c**, a modified augmented **fourth- sixth chord.**

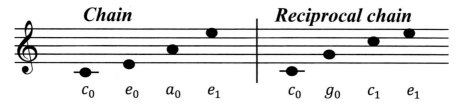

Since the chain $2 : 3 : 4 : 5$ is obviously arithmetic, its reciprocals must be harmonic according to the theory we saw in b) by verifying.   ◄

The next example combines the **dominant seventh chord** with a **sixth chord in the ecclesiastical mode "dorian"**:

Example 5.9 Four-tone chords and Their Chains of Proportions

Now three musical intervals, given by virtue of the proportions

$$a : b \cong 8 : 9, \; b : c \cong 4 : 5 \text{ and } \; c : d \cong 5 : 6$$

and let it be $a : b : c : d$ the assembled proportion chain of a four-tone chord. We set ourselves the **task** again:

(a) Describe the chord musically.
(b) Calculate the matching number proportion chain.
(c) Construct the reciprocal chain and describe a chord that corresponds to this reciprocal.

The solution (to (a)): This time we start with the note f, which is followed by the note g after the whole tone step $8 : 9$; this is followed by the sub-chain $(4 : 5) \oplus (5 : 6)$, a true major chord in just diatonic temperament. All in all, we then hear an inversion of the **dominant seventh chord** g^7

$$f - g - h - d^1.$$

We obtain the chain of proportions by adjunction of the sub-chain $(4 : 5 : 6)$ to the proportion $(8 : 9)$, which is done by extending it by 9 or by 4, respectively; then we obtain the chain which cannot be shortened further

$$(32 : 36) \oplus (36 : 45 : 54) \cong 32 : 36 : 45 : 54$$

of the total proportion of a sixth $16 : 27$ in pythagorean temperament, which also takes care of part (b). To solve part (c), we again proceed "musically". We reverse the dominant seventh chord by arranging the proportions reciprocally; then we get the adjunction process

$$(5 : 6) \oplus (4 : 5) \oplus (8 : 9) \cong (10 : 12 : 15) \oplus (8 : 9) \cong 80 : 96 : 120 : 135,$$

which of course has the same outer proportion $16 : 27$. A chord played on white keys of a diatonically just tempered instrument would be (only) the **D minor chord** with the **major sixth ('sext ajoutée – dorian sixth')**,

$$d - f - a - h.$$

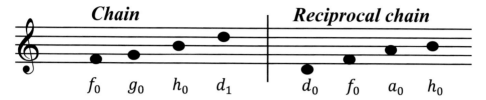

It is therefore the inversion of the dominant seventh chord when it begins on the seventh of the dominant. ◀

These contexts can be wrapped up in numerous other tasks.

5.4 Symmetries in the Interval Cosmos: Geometric Chains of Proportions and the Harmonic Equations

In music theory, the geometric chains are probably the most prominent of the mean value chains. This is primarily because there is a significant connection between the frequency measure proportions and the construction of **regularly layered** interval structures (such as scales, chords). The theorem that now follows is ultimately a kind of background mathematics for some of these relations that can be described in concrete situations – but which can perhaps be understood more sustainably in the light of this background.

Theorem 5.1 (Interval Adjoint Functions and the Proportion Chains of Their Measures)

Let there be $I_1, \ldots, I_n$ musical intervals to the proportions $A_1 = a_1 : b_1, \ldots, A_n = a_n : b_n$. Furthermore let the proportions

$$X_k \cong A_{k+1} \odot A_k^{\text{inv}}, \quad k = 1, \ldots, n-1$$

be the differences – i.e. the solutions of the equation that are unique except for similarity

$$X_k \odot A_k \cong A_{k+1}, \quad k = 1, \ldots, n-1.$$

Then the following four statements are equivalent:

1. The proportion chain formed from the **frequency measures of** the intervals

(continued)

Theorem 5.1 (continued)

$$\left(\frac{b_1}{a_1}\right) : \left(\frac{b_2}{a_2}\right) : \dots : \left(\frac{b_n}{a_n}\right) \quad \text{respectively} \quad \mid I_1 \mid : \mid I_2 \mid : \dots : \mid I_n \mid$$

is a geometric chain of proportions.

2. The proportion chain formed from the **cent measures** of the intervals

$$ct(I_1) : ct(I_2) : \dots : ct(I_n)$$

is a arithmetic chain of proportions.

3. The differences $J = I_{k+1} \ominus I_k$ of adjacent intervals are always the same interval J for all stages $k = 1, \dots, n - 1$, and the interval series forms an iteration of adjuncts of this **iteration interval** J to the starting interval I_1; in detail:

$$I_1 = \text{start}, I_2 = I_1 \oplus J, I_3 = I_1 \oplus 2J, \dots, I_n = I_1 \oplus (n-1)J.$$

4. The $(n-1)$-step proportion chain of adjoint differential proportions

$$X_1 \oplus \dots \oplus X_{n-1}$$

is a geometric chain of proportions.

The justification of the theorem happens roughly like this: First of all, statements (1.) and (2.) are equivalent, since this is a property of the logarithm:

Mathematical Background   The logarithm of a product is the sum of the logarithms of the individual factors.

We show the equivalence (1.) $\Leftrightarrow$ (3.): To do this, we pick out three consequent magnitudes of the proportion chain from (1.)-without limitation, let this be the 3-membered chain

$$\left(\frac{b_1}{a_1}\right) : \left(\frac{b_2}{a_2}\right) : \left(\frac{b_3}{a_3}\right) = \alpha_1 : \alpha_2 : \alpha_3.$$

By definition, it is geometric if and only if $\alpha_2 / \alpha_1 = \alpha_3 / \alpha_2$, and then the chain $\alpha_1 : \alpha_2 : \alpha_3$ with the step parameter $\lambda = \alpha_2/\alpha_1 = \alpha_3/\alpha_2$ has the form of the geometric sequence

$$\alpha_1 : \lambda\alpha_1 : \lambda^2\alpha_1.$$

Now if J is the interval whose frequency measure has exactly the value λ – symbolically

$$J = [1, \lambda] = [\alpha_1, \alpha_2] = [\alpha_2, \alpha_3],$$

so is $I_2 = I_1 \oplus J$ and $I_3 = I_1 \oplus 2J$ since the frequency measure of a sum (adjunction) of several intervals is the product of all frequency measures. The equivalence (3.) $\Leftrightarrow$ (4.) is trivial and our theorem is proved.

The following four examples deal with the relationships in the just diatonic temperament – that is, in the interval systems built up by major thirds 4 : 5 and perfect fifths 2 : 3 as well as octaves the system $\mathbb{P}_{\text{diat}}$. First, we find correlations in the three semitone intervals (semitonia) of the **diatonic,**

diatonic semitone, pythagorean limma and Eulerian semitone.

Then follows an interesting aspect for the three semitonia of **chromaticism,**

small and large chroma and pythagorean apotome.

These are not actual diatonic - heptatonic intervals but which are calculated as complementary intervals of the diatonic semitones in whole tones. Therefore, they occur in the 12-step 'chromatic' scales (see for example the detailed literature in [16]). The differences among these diverse semitonia lead to the 'quarter tone' intervals (microtonal intervals, commas) and we recognize regularities in the three Commas of **enharmonics:**

syntonic comma, pythagorean comma and diaschism.

These three microintervals form surprisingly remarkable proportion chains, as do the three commas **diaschism, lesser** and **greater diësis.**

▶   In all four cases, remarkable orders emerge: All three intervals form **geometric** chains of proportions. These relations are called **'harmonic equations'.**

A short remark about the notation: For the frequency measure of an interval I we use the symbol '|name of the interval|' as we have already done many times.

Example 5.10 Geometric Proportion Chains in Diatonics

The semitonia (ordered by size) of the just-tuned chromatic system

Pythagorean limma (L) – diatonic semitone(S) – Euler semitone (E)

form – in their frequency factors – a geometric proportion chain. The difference in the next larger interval is in each case a syntonic comma and we obtain the 'harmonic equations':

Harmonic equations for the semitonia of diatonics

1. $|\text{Limma}| : |\text{diatonic semitone}| \cong |\text{diatonic semitone}| : |\text{Euler-semitone}|$,

2. diatonic semitone $=$ Limma $\oplus$ syntonic comma,

3. Euler $-$ semitone $=$ diatonic semitone $\oplus$ syntonic comma.

Why? Thanks to the equivalences of Theorem 5.1 we have several possibilities to show this:

(A) The calculation of the frequency factor analysis shows concretely the proportion values

$$\frac{256}{243} : \frac{16}{15} = \frac{80}{81} = \frac{16}{15} : \frac{27}{25}$$

which gives a geometric chain of proportions.

(B) Interval arithmetic: These three semitone steps are constructed like this:

Pythagorean Limma $\oplus$ syntonic comma $=$ diatonic semitone,
diatonic semitone $\oplus$ syntonic comma $=$ Euler-semitone;

here a corresponding definition of this semitonia can be taken as a basis but we do not pursue this further in the context of this consideration.

(C) The difference proportions are

$$X_1 = (16 : 15) \odot (243 : 256) = (16 * 243) : (15 * 256) \cong 81 : 80,$$
$$X_2 = (27 : 25) \odot (15 : 16) = (27 * 15) : (25 * 16) \cong 81 : 80,$$

which means that they are equal (similar) and $X_1 \oplus X_2$ is a geometric chain.   ◀

Conclusion 'The diatonic semitone (S) is musically the **geometric mean of** the limma (L) and Eulerian semitone E' – these three form a geometric proportion chain! The step proportion is the syntonic comma and we formulate this in the catchy symbolic verbal formula:

Symbolic Proportion Equation of Diatonics

$$\text{diatonic semitone} = \frac{1}{2}\left(\text{Pythagorean Limma} \oplus \text{Euler-semitone}\right)$$

The next example connects the 'chromatic' semitones:

Example 5.11 Geometric Proportion Chains in Chromatics

The semitones (ordered by size) of the just-tuned chromatic system

$$\text{small chroma}(\text{ch}) - \text{large chroma}(\text{CH}) - \text{Pythagorean apotome (A)}$$

form – in their frequency factors – a geometric proportion chain. The difference in the next larger interval is in each case a syntonic comma. In formulas:

Harmonic equations for the semitonia of chromaticism

1. $|\text{small chroma}| : |\text{large chroma}| \cong |\text{large chroma}| : |\text{apotome}|$,

2. $\text{large chroma} = \text{small chroma} \oplus \text{syntonic comma}$,

3. $\text{apotome} = \text{large chroma} \oplus \text{syntonic comma}$.

Why? We make this clear by showing that the square of the frequency measure of the large chroma is equal to the product of the other two semitones – this is, after all, the shortest provable characterization for the geometric mean. By comparing the prime number exponents we see immediately – that is without calculating – the equality

$$\left(3^3 5^1/2^7\right)^2 = \left(5^2/2^3 3^1\right) * \left(3^7/2^{11}\right).$$

The difference – i.e. the quotient CH/ch of the measures– then quickly leads to the frequency measure of the syntonic comma 81/80.   ◀

Conclusion 'The large chroma is the geometric mean of the small chroma and the pythagorean apotome' and step-proportion is (again) the syntonic comma – there is another symbolic equation:

Symbolic Proportion Equation of Chromatics

$$\text{large Chroma} = \frac{1}{2}(\text{small Chroma} \oplus \text{Pythagorean Apotome})$$

Once again, this relationship could be equivalently reformulated in the numerical data.

The third example takes us into the world of microintervals, specifically to **commas**, which are named like this if they describe certain **balance deficits**, such as

- **pythagorean comma:** 12 pythagorean fifths against 7 octaves,
- **lesser diësis:** 1 octave against 3 just major thirds,
- **greater diësis:** 4 just minor thirds against an octave,
- **syntonic comma:** 4 fifths minus 2 octaves against 1 just major third,
- **diaschism:** 3 octaves versus the sum of 4 fifths and 2 just major thirds,

to name only the best known and most important ones. Finally, the tiny **schism** is the difference between pythagorean and syntonic comma. Thus it presents itself through the – the diatonic generating intervals – in the balance

- **schism:** 8 fifths minus 5 octaves versus 1 third.

and it has the tiny size of ≈ 2 ct.

Example 5.12 Geometric Proportion Chains in the Enharmonics of Commas

The commas (ordered by size)

$$\text{Diaschisma} - \text{Syntonic comma} - \text{Pythagorean comma}$$

form – in their frequency factors – a geometric proportion chain. The difference in the next larger interval is in each case a **schism.** In formulas:

Harmonic equations for the commas of the enharmonics

1. $|\text{diaschisma}| : |\text{syntonic comma}| \cong |\text{syntonic comma}|$
   $: |\text{pythagorean comma}|$,

2. syntonic comma $=$ diaschism $\oplus$ schism,

3. pythagorean comma $=$ syntonic comma $\oplus$ schism.

Why? Here we content ourselves with the proof that both difference proportions are similar:

$$X_1 = \left(2^4 5^1 : 3^4\right) \odot \left(2^{11} : 3^4 5^2\right) = 2^{15} 5^1 : 3^8 5^2 \cong 2^{15} : 3^8 5^1,$$
$$X_2 = \left(2^{19} : 3^{12}\right) \odot \left(3^4 : 2^4 5^1\right) = 2^{19} 3^4 : 2^4 3^{12} 5^1 \cong 2^{15} : 3^8 5^1,$$

and both times the same proportion is produced – namely that of the tiny microtone **schism.** For your own practice we recommend to do the proof once by means of a frequency-factor-proportion-chain.   ◀

Conclusion The syntonic comma is the geometric mean of the diaschism and the pythagorean comma, the step proportion is the schism – we get another symbolic equation:

Symbolic Comma Proportion Equation of Enharmonics

$$\text{syntonic Comma} = \frac{1}{2}\left(\text{Diaschism} \oplus \text{Pythagorean Comma}\right)$$

with which we have also found a geometric mean formula for this case.

In the realm of microtones, the two dièses are especially famous – after all, they measure the deviations of just minor or major thirds from the octave; it is precisely because of these balances that they have become indispensable in the **theory of temperaments** – that is, in the scale theory of the Bach-age. On the contrary: here their practical significance probably goes far beyond that of the two commas of antiquity (pythagorean and syntonic comma).

Example 5.13 Geometric Proportion Chains in the Enharmonics of the Diëses
The commas (ordered by size) of the perfect diatonics

Diaschism – lesser Diësis – greater Diësis

form – in their frequency factors – a geometric proportion chain. The difference in the next larger interval is in each case a **syntonic comma,** in formulas:
   Harmonic equations for the diëses of the enharmonics

1. $|\text{lesser diësis}| : |\text{diaschisma}| \cong |\text{greater diësis}| : |\text{lesser diësis}|$,

2. lesser diësis $=$ diaschism $\oplus$ syntonic comma,

3. greater diësis $=$ lesser diësis $\oplus$ syntonic comma.

Why? We want to consult the cent measures (see table of the appendix). Accordingly, we read off the values:

$$\text{ct (diaschism)} = 19,5 \text{ ct,}$$
$$\text{ct (lesser diësis)} = 41,0 \text{ ct,}$$
$$\text{ct (greater diësis)} = 62,5 \text{ ct.}$$

The cent differences are thus both times 21.5 ct – which is the cent measure of the syntonic comma. By applying Theorem 5.1, according to which these cent measures thus represent an arithmetic series, we thus arrive at the statement above.

First, however, a small downer must be noted because unfortunately this calculation is not 'strictly proving' since all these cent numbers are rounded – there could still be a difference in a more distant decimal place.

Though it doesn't as the (exact) frequency measure quotients prove, and when we set $X =$ (lesser diësis) minus (diaschism) and $Y =$ (greater diësis) minus (lesser diësis), then the calculation holds:

$$X = \left(2^7 5^{-3}\right) / \left(2^{11} 3^{-4} 5^{-2}\right) = 2^{-4} 3^4 5^{-1}$$
$$Y = \left(2^3 3^4 5^{-4}\right) / \left(2^7 5^{-3}\right) = 2^{-4} 3^4 5^{-1}.$$

The same quotients arise both times – these are apparently the exact frequency measure of the syntonic comma.   ◀

Conclusion The lesser diësis is the geometric mean of the diaschism and the greater diësis, the step proportion is the syntonic comma – we get another symbolic equation:

Symbolic Diësis Proportion Equation of Enharmonics

$$\text{lesser Diësis} = \frac{1}{2}\left(\text{Diaschism} \oplus \text{greater Diësis}\right)$$

It describes symbolically as well as numerically the intrinsic interrelation of these microtonal families.

We could certainly find quite a few more examples with a good chance of success; the table in the appendix is a call to do so.

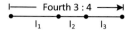

Fig. 5.2 The tetrachord model

5.5   Tetrachordics: Applied Musical Proportion Theory

Anyone who delves into ancient – predominantly greek – music theory will first of all discover a respectable abundance of interval structures – especially when these are combined in a 'triplets package' to form the tonal space of a fourth. The result is a tetrachord. Tetrachords were used to build the early scale scales and the names '**dorian**', '**phrygian**', '**lydian**', and several others can be explained from the architecture of these tetrachords and their use. Table 5.3 certainly gives us a first impression.

Answer and Question '*Thank God there is nevertheless an – albeit rough – order in the apparent confusion of these – mostly by proportions characterized – structural elements*'. *How do we come to such a helpful clear order in this musical world?*

Within the scope of our possibilities, however, we can only give an outline of an order for these structures; the cross-connections to the theory of proportions are our guidelines. We begin with the definition of a 'tetrachord':

A tetrachord is a scale made up of exactly 3 upward intervals I_1, I_2, and I_3 – or a sequence of 4 ascending tones – of the total range of a just (pythagorean) fourth 3 : 4 as shown in Fig. 5.2.

The step intervals I_1, I_2, and I_3 can be very different and can sometimes be located in the micro- or quarter-tone range.

Usually, tetrachords have been classified by a two-parameter 3 × 3 grid of characterization features. This is done by dividing them into

(A) three harmonious genders:
      diatonic – chromatic – enharmonic,
(B) three families:
      dorian – phrygian – lydian

having the following meaning:

(1) The **gender** indicates which types of steps (intervals I_1, I_2, I_3) occur in the tetrachord, whereby an approximate but variable division of these steps allows assignment to the following basic types (the (variable) symbol indications refer to Table 5.2):
     • quarter tones – microtones – (symbol µ),
     • semitones – (symbol S),

Table 5.2 Basic patterns of tetrachordics. (After Euclid and Nicomachus)

A	B Phrygian			Dorian			Lydian		
Enharmonic	μ	μ	T + T	T + T	μ	μ	μ	T + T	μ
Chromatic	S	S	T + S	T + S	S	S	S	T + S	S
Diatonic	S	T	T	T	S	T	T	T	S

- whole tone (symbol T),
- minor and major third (symbols $T + S$ and $T + T$).

These basic types can actually also vary greatly among themselves and this (unfortunately) has a strong arbitrary character. The tone genders are then determined solely by the **occurrence of these basic types** – as follows:

- **diatonic gender** ↔ tonos, tonos, semitonos;
- **chromatic gender** ↔ minor third, semitone, semitone;
- **enharmonic gender** ↔ major third, quarter tone, quarter tone.

(2) The three **families** (genera) **'dorian', 'phrygian', 'lydian'** determine the **order** in which these intervals from the gender (A) are arranged in the tetrachord, whereby essentially the position of the smallest and/or the largest interval becomes the characteristic.

Thus there are 9 types of tetrachords and Table 5.2 describes in a suitable overview the classification of the tetrachords according to these two characteristics – in the general case of different whole tones; the interval series are given in the upward direction (contrary to the ancient listing, which corresponded to the downward melodies). Whereby, unfortunately, a uniformity of this description was not and is not given either in antiquity or in contemporary scholarship. In Table 5.2 we follow the conception of Thimus, who again relies on sources around Euclid, Nicomachus and Boethius (but also contradicts them).

In relation with the teaching of the ancient greek, the gregorian, and the ecclesiastical modes, the **diatonic** gender is now exclusively of importance. It is easier to give a clear characterizing description of the families:

- **phrygian diatonic** ↔ semitonium is the **front step of** the tetrachord,
- **dorian-diatonic** ↔ semitonium is the **middle degree of** the tetrachord,
- **lydian-diatonic** ↔ semitonium is the **back step of** the tetrachord.

At this point we must note that – unfortunately and very often – both families 'dorian' and 'phrygian' can appear in reversed roles. What is dorian in one case, might be phrygian in another case and vice versa.

Thus, for example, in some lexicons the ancient greek scale $e_0 - e_1$ is called 'dorian'; the dorian 'ecclesiastical mode', on the other hand, is the scale $d_0 - d_1$ and for the term 'phrygian' the opposite is accurate (the sequence of tones belongs to the 'white keyboard keys').

The reasons for this are manifold, controversial, and they are fed by different views on the background of the historical principles which the **'systema teleion'** or **'systema maxima'** used as a perfect tone and tetrachord system for the development of all tonalities. The term 'lydian', on the other hand, is not subject to such ambivalence. This difference is most noticeable when ecclesiastical modes are compared to ancient greek modes and for this purpose exist the following illustrious

Interjection *The french composer Jehan* **Alain** *(1911–1940) left a striking oevre pour l'orgue. His two organ works*
  Choral dorien [1935] and Choral phrygien [1935].
  belong to the most popular and frequently played pieces of the french organ literature.
  The only difference is that the **'choral dorien'** *is not dorian, but phrygian – namely in the gregorian mode of tonus IV, and conversely the* **'choral phrygien'** *is not phrygian, but dorian in the tonus gregorianus II (the differences ultimately stem from erroneous sources of earlier late medieval literatures).*

Table 5.3 gives an impression of the richness of tetrachordal structures. It is worthwhile to construct one or the other proportion chain and to examine it with regard to its arithmetic or harmonic medietary patterns.

Thus the proportion chain of the **'dorian chroma of Eratosthenes'** is the sequence

$$15 : 18 : 19 : 20;$$

it contains the arithmetic sub-chain 18 : 19 : 20 with new ecmelic semitones of the difference intervals – due to the prime number 19.

All of these terms translate directly to the corresponding proportion chains, and we also recognize certain symmetries between a tetrachordal proportion chain and its reciprocal:

▶   For **enharmonic and chromatic** tetrachord-chains, the 'dorian' and 'phrygian' are reciprocal, and the 'lydian' property is preserved under reciprocity. In contrast, for **diatonic** tetrachord-chains, the 'phrygian' and 'lydian' are reciprocal and the 'dorian' property is preserved under reciprocity.

The following examples make use of this classification while combining the tetrachord theory with the proportion theory. We begin with the

Example 5.14 Pythagorean Diatonic Tetrachords

Pythagorean diatonic tetrachords are constructed exclusively of the single whole tone type tonos 8 : 9 and consequently of the semitone limma 243 : 256. Therefore, the three families can be described:

Table 5.3 **Table** of proportions of some ancient Greek tetrachords

Tetrachord type	Proportion dimensions of the steps			Cent measurements (rounded)		
Archytas of Tarent (fourth century BC)						
Dorian enharmonic	4:5	35:36	27:28	386	49	63
Dorian chroma	27:32	224:243	27:28	294	141	63
Phrygian diatonic	15:16	8:9	9:10	112	204	182
Lydian diatonic	8:9	7:8	27:28	204	231	63
Erathostenes (third century BC)						
Dorian enharmonic	15:19	38:39	39:40	409	45	44
Dorian chroma	5:6	18:19	19:20	316	93	89
Lydian diatonic	8:9	8:9	243:256	204	204	90
Didymos (first century BC)						
Dorian enharmonic	4:5	30:31	31:32	386	57	55
Dorian chroma	5:6	24:25	15:16	316	71	112
Lydian diatonic	8:9	9:10	15:16	204	182	112
Ptolemy (first century AD)						
Dorian enharmonic	4:5	23:24	45:46	386	74	38
Dorian soft chroma	5:6	14:15	27:28	316	119	63
Dorian hard chroma	6:7	11:12	21:22	266	151	81
Lydian soft diatonic	7:8	9:10	20:21	231	182	84
Lydian whole tone diatonic	8:9	7:8	27:28	204	231	63
Lydian pythagorean. diatonic	8:9	8:9	243:256	204	204	90
Lydian hard diatonic	9:10	8:9	15:16	182	204	112
Lydian evenly diatonic	9:10	9:10	25:27	182	182	134

(A) The **lydian-pythagorean-diatonic tetrachord** carries the classification pattern $1 - 1 - 1/2$, and we therefore have the proportion chain

$$[8, 9] \oplus [8, 9] \oplus [243, 256] \rightleftarrows 192 : 216 : 243 : 256.$$

(B) The **phrygian-pythagorean-diatonic tetrachord** carries the classification pattern $1/2 - 1 - 1$, and we therefore have the proportional chain

$$[243, 256] \oplus [8, 9] \oplus [8, 9] \rightleftarrows 243 : 256 : 288 : 324.$$

(C) The **dorian-pythagorean diatonic tetrachord** carries the classification pattern $1 - 1/2 - 1$, and we therefore have the proportion chain

$$[8, 9] \oplus [243, 256] \oplus [8, 9] \rightleftarrows 216 : 243 : 256 : 288.$$

We immediately see the symmetries, that the lydian and phrygian are reciprocal to each other and that the dorian tetrachords are symmetrical – if we look at all this through the "glasses of proportion".   ◀

We follow this example with the important component of the diatonic:

Example 5.15 Lydian and Phrygian Diatonic Tetrachords of Archytas

(A) The **lydian** diatonic **tetrachord of Archytas** bears the classification pattern $1 - 1 - 1/2$, as a characteristic of the diatonic gender of the lydian form and it has the more precise interval structure or proportion chain

$$[8,9] \oplus [9,10] \oplus [15,16] \rightleftarrows 24 : 27 : 30 : 32.$$

This would correspond, for example, to the tone sequence $c - d - e - f$ with the more precise (!) steps

(pythagorean whole tone) $\oplus$ (diatonic whole tone) $\oplus$ (diatonic semitone).

If we swap both whole tone steps, we get another almost identical appearing lydian-diatonic tetrachord

$$[9,10] \oplus [8,9] \oplus [15,16] \rightleftarrows 36 : 40 : 45 : 48.$$

Both tetrachords – connected by a pythagorean whole-tone step – also define the important just-tuned **diatonic heptatonic octave scale.** In Example 5.4, we have already presented the proportions of this scale.

(B) A reciprocal of this lydian-diatonic tetrachord with the consequently also reciprocal classification pattern $1/2 - 1 - 1$ is now the **phrygian-diatonic tetrachord of Archytas,**

(diatonic semitone) $\oplus$ (diatonic whole tone) $\oplus$ (pythagorean whole tone).

However, it could not be played on the same white keyboard-keys as the lydian diatonic without swapping the two whole tone steps (WTS) because there is no sequence of upward steps 'diatonic WTS– pythagorean WTS' implanted.
The proportion chain of this dorian diatonic tetrachord is the composition

$$(15 : 16) \oplus (9 : 10) \oplus (8 : 9) \cong (135 : 144) \oplus (144 : 160) \oplus (8 : 9)$$
$$\cong (135 : 144 : 160) \oplus (160 : 180) \cong 135 : 144 : 160 : 180.$$
  ◀

In the next example we see a clearly different tetrachord, which nevertheless has the classification lydian-diatonic:

Example 5.16 Lydian and Phrygian Diatonic Tetrachords of Ptolemy

(A) The lydian diatonic tetrachord of Ptolemy has the following structure:
The ecmelic whole tone 7 : 8 is followed by the just diatonic whole tone 9 : 10 and a small semitone 20 : 21 (about 84, 5 ct), which guarantees the balance to the just fourth.

$$[7, 8] \oplus [9, 10] \oplus [20, 21] \rightleftarrows 63 : 72 : 80 : 84.$$

(B) The reciprocal chain provides Ptolemy's phrygian diatonic tetrachord

$$[20, 21] \oplus [9, 10] \oplus [7, 8] \rightleftarrows 60 : 63 : 70 : 80.$$

Here, too, the fastest way to calculate the proportion chain is to expand the respective internal magnitudes step by step.   ◀

In the next part we give an example of a phrygian-chromatic – hence a non-diatonic – tetrachord – and its reciprocal partner, a dorian-chromatic tetrachord.

Example 5.17 Dorian and Phrygian Chromatic Tetrachords of Didymos

(A) We start with a tetrachord that goes back to Didymos (first century BC):
The architecture is shown in the following step proportions with the corresponding proportion chain:

$$(5 : 6) \oplus (24 : 25) \oplus (15 : 16) \rightleftarrows 60 : 72 : 75 : 80.$$

We recognize familiar intervals; it has the structure

(just minor third) ⊕ (small chroma) ⊕ (diatonic semitone),

and the entire proportion chain belongs to the just-tuned-diatonic interval circle of the Senarius. All intervals are simple superparticular. The small chroma is the complementary interval of the diatonic semitone in the minor whole tone – consequently it belongs to the class of 'semitones'. Consequently, the tetrachord has the classificatory structure

$$3/2 - 1/2 - 1/2$$

and thus justifies the name '**dorian chromatic tetrachord' of Didymos.**

(B) The tetrachord reciprocal to this will thus be the proportion structure

$$(15 : 16) \oplus (24 : 25) \oplus (5 : 6) \rightleftarrows 45 : 48 : 50 : 60$$

and is, according to classification, a **phrygian-chromatic tetrachord of Didymos.**   ◀

Finally, an example from the family of enharmonic tetrachords should follow and for this we have a tetrachord with rather strange proportions, namely the tetrachord of Eratosthenes (ca. 296–194 BC):

Example 5.18 Dorian and Phrygian Enharmonic Tetrachords by Eratosthenes

(A) The following construct goes back to Eratosthenes: the microtonal structure of his **dorian-enharmonic tetrachord** has the proportions

$$(15 : 19) \oplus (38 : 39) \oplus (39 : 40) \rightleftarrows 30 : 38 : 39 : 40.$$

There are no more 'familiar' intervals of the Senarius; after all we recognize that the generating prime factors besides 2, 3, and 5 are also 13 and at the end even 19. The cent numbers outlin steps are these:

$$409, 2 \text{ ct} - 45, 0 \text{ ct} - 43, 8 \text{ ct},$$

and the interval $(15 : 19)$ with its $409, 2$ ct is thus almost identical to the ditonos, the pythagorean major third $(64 : 81)$ with $407, 8$ ct.
How close these two intervals are to each other can also be seen from their proportions, independently of the cent numbers, which are probably not easy to calculate. For this purpose we build both proportions on a uniform magnitude. Then

$$15 : 19 \cong 960 : 1216 \text{ and } 64 : 81 \cong 960 : 1215,$$

is namely a hardly measurable (as well as audible) difference. They are actually 'the same'.

(B) The reciprocal inversion leads to the **phrygian-enharmonic tetrachord**

$$(39 : 40) \oplus (38 : 39) \oplus (15 : 19) \; \rightleftarrows \; 741 : 760 : 780 : 988,$$

whose proportion chain has large numerical values, as expected, since in this case the connecting magnitudes are unfavourably matched.   ◄

In the next example, we encounter a rather interesting tetrachord: as we will see, it has an exceptional character in many respects.

We may ask ourselves what the proportions of a tetrachord would look like if the outer proportion – i.e. the fourth 3 : 4 – were divided harmonically and arithmetically ('musically'). Since, in the case of the harmonic mean in particular, the total interval is then itself to be divided again in the ratio 3 : 4, whereas in the case of the arithmetic mean it is only to be halved in the ratio 1 : 1, we come – however – to the idea of using a priori the proportion 42 : 56 that is similar to 3 : 4. Afterwards we actually have the integer medieties 49 (arithmetic mean of 42 and 56) and 48 (harmonic mean of 42 and 56), and we get the mean value chain in the form of a Harmonia Perfecta Maxima, namely

$$42 : 48(=y_{\text{harm}}) : 49(=x_{\text{arith}}) : 56,$$

which defines the proportion chain of our targeted tetrachord. This sequence of numbers thus reflects the famous musical **proportion of Iamblichos**, of which we also know that the first and third proportions are equal (that means: similar).

Example 5.19 The Tetrachord to the Musical Proportion Chain

The arithmetic-harmonic division of the fourth results in the proportion chain of the babylonian canon $P_{\text{mus}}(3 : 4)$ according to the laws of our harmonia perfecta maxima babylonica from Sect. 3.2 and it is simultaneously a tetrachord with the structure of composition

$$42 : 48 : 49 : 56 \; \rightleftarrows \; (7 : 8) \oplus (48 : 49) \oplus (7 : 8).$$

This tetrachord is the **tetrachord to the musical proportion chain of Iamblichos** – or **'tetrachord of harmonic and arithmetic medieties'**. It has the following properties:

1. According to our classification, it is a dorian-diatonic tetrachord.
2. The two augmented, but equally large whole tone steps are the ecmelic whole tones 7 : 8 (with $\approx 231,2$ ct), formed with the prime number 7.
3. Therefore, the 'semitone step' 48 : 49 (with approximately 35, 7 ct) is left with only a microtonal character, and could be categorized as a 'close quarter tone'.
4. The tetrachord is symmetrical and therefore identical to its reciprocal form.
5. All intervals are simple-superparticular – that is, the proportions are of the form $n : (n + 1)$; the numerator of the frequency measure fraction is 1 greater than the

denominator. Such intervals were especially important in antiquity, since (in many cases) only those were considered '**consonant**'.

6. You can show (though with some effort):

> **Theorem:** There is exactly one diatonic tetrachord with exclusively single-superparticular steps, in which the two whole-tone steps are of equal size: this is the present tetrachord to the musical proportional chain (cf. [16], Sect. 4.7 ◄

It would also be interesting to ask what a tetrachord would look like if the interval of fourths 3 : 4 were divided by the two medieties '**contra-arithmetic**' and '**contra-har-monic**'. According to Nicomachus' theorem or according to the harmonia perfecta maxima diatonica (Theorem 3.8), a symmetrical tetrachord would again have to be created – that is, a tetrachord whose first and third steps are similar. After all, the symmetry rules of our Theorem 3.8 apply together with the cross-rule:

$$a : x_{\text{contra-arith}} \cong y_{\text{contra-harm}} : b \Leftrightarrow a : y_{\text{contra-harm}} \cong x_{\text{contra-arith}} : b.$$

For the magnitudes $a = 42$ and $b = 56$ for the fourth chosen in Example 5.19, using our mean value formulas of Theorem 3.3, we would obtain the values

$$x_{\text{contra-arith}} = 47,04 \text{ and } y_{\text{contra-harm}} = 50.$$

The desire for integerity leads (after extension with 25/2) to a representation that can no longer be simplified:

$$42 : y_{\text{contra-harm}} : x_{\text{contra-arith}} : 56 \cong 525 : 588 : 625 : 700.$$

This is the **tetrachord of the contra-harmonic and contra-arithmetic medieties.** The cent measures of the steps yield the values

$$196, 2\,\text{ct} - 105, 6\,\text{ct} - 196, 2\,\text{ct}.$$

Interesting: These steps are therefore not very far from those of the equal-tuned tempera-ture, here it would be the smooth values

$$200\,\text{ct} - 100\,\text{ct} - 200\,\text{ct}.$$

According to classification, this tetrachord is also a dorian diatonic tetrachord.

Our last example shows two tetrachords in the peculiar situation of almost equal interval steps; they gradually diminish in an arithmetic manner: while one corresponds to an arithmetic proportion chain, in the other – the reciprocal – the proportion chain is conse-quently of harmonic type.

Example 5.20 Maramures Tetrachord: Arithmetic and Harmonic Tetrachord

(A) If one defines a tetrachord by the following remarkable arrangement

$$[9, 10] \oplus [10, 11] \oplus [11, 12] \rightleftarrows 9 : 10 : 11 : 12,$$

it becomes clear that the chain of proportions is an arithmetic chain of proportions. A simple consideration shows that this is the only way to find an arithmetic 3-step proportion chain of the total proportion of a fourth. The sequence staggered in this way is called a '**maramures chain**'. The logarithmic measures of the step intervals are

$$182, 4\,\text{ct} - 165, 0\,\text{ct} - 150, 6\,\text{ct}.$$

(B) According to our Theorem 3.5, a reciprocal chain of proportions must be harmonic; we once calculate the proportions of the inverse arrangement of the chain

$$(11 : 12) \oplus (10 : 11) \oplus (9 : 10) \cong (55 : 60) \oplus (60 : 66) \oplus (9 : 10)$$
$$\cong (55 : 60 : 66) \oplus (9 : 10) \cong (165 : 180 : 198) \oplus (198 : 220)$$
$$\cong 165 : 180 : 198 : 220,$$

and this expression cannot be further simplified by shortening to integer. Nevertheless, we see without effort that the chain is harmonic. For 180 is the harmonic mean of 165 and 198 – it is, after all, by application of the definition

$$(180 - 165) : (198 - 180) = 15 : 18 \cong (11 * 15) : (11 * 18) \cong 165 : 198,$$

and likewise 198 is also the harmonic mean of 180 and 220 – for

$$(198 - 180) : (220 - 198) = 18 : 22 \cong 180 : 220.$$

Certainly, both tetrachords elude classification into the ancient grid of the 9 main types of tetrachords (genus and family) – but this does not diminish their charm:
Theorem: The tetrachord 9 : 10 : 11 : 12 is the only tetrachord with an arithmetic chain of proportions, and consequently the tetrachord

$$165 : 180 : 198 : 220$$

also the only one with a harmonic chain of proportions.

While the first of these two statements of the Theorem is certainly trivial in nature, the proof of the second statement – concerning the harmonic chain – would hardly succeed without major effort, were it not for the theoretical general symmetry property of our central Theorem 3.5! ◄

Conclusion Greek tetrachordics provides countless further examples – some of them of a bizarre nature – whose systematic reappraisal would certainly provide worthwhile insights into the diversity of scales and interval structures. And here, too, the theory of the symmetries of the mean value proportion chains – the harmonia perfecta – would undoubtedly be of great use.

5.6 Modology: Gregorian Mathematics

The study of the ancient keys is a science of which whole libraries know to report. If the richness of tetrachordics already led to dozens of proportion chains – how much more should this be true for heptatonic scales, which one has to imagine as being composed of tetrachords (and a connecting major whole tone). Thus we cannot attempt to provide satisfactory information on this historically and substantively extremely extensive subject; too many names, terms and historical events would await mention and explanation. However, we would like to offer at least an introductory overview in order to present at least some basic types of **church tonal scales** in their standard proportion chains. We also want to outline, in the required brevity, how everything – perhaps -came into being at some time and somehow.

A description of ancient greek scales is based on the **systema teleion** (or **systema maximum immutabile**) – a two-octave system 'containing all tones' – which is itself composed of four downward tetrachords.

In the context of this systema teleion and the structure of both the ancient greek and its kindred church tonal scales, these tetrachords are exclusively of the **diatonic** gender. Since the systema teleion comprises two octaves, it is also sufficient that this system is formed exclusively by **phrygian downward tetrachords.** For it turns out that in their combinations to octochords – i.e. in heptatonic octave scales – both dorian and lydian tetrachords are formed.

The diatonic tetrachord is determined by two whole tones (which can be different) and a semitone that fills the difference to the fourth. There are two types of whole tones in particular which come to the fore: These are the two whole-tone steps 8 : 9 and 9 : 10, and formally the following – in the lydian variant – written constellations would result

1. $[8, 9] \oplus [8, 9] \oplus [243, 256]$,

2. $[9, 10] \oplus [9, 10] \oplus [25, 27]$,

3. $[8, 9] \oplus [9, 10] \oplus [15, 16]$ or $[9, 10] \oplus [8, 9] \oplus [15, 16]$.

The three resulting semitonia are the already known intervals

$$\text{Pythagorean Limma } L \ (243 : 256),$$
$$\text{Diatonic Semitone } S \ (15 : 16)$$

and the less common

$$\text{Euler Semitone } E \ (25 : 27),$$

which we saw in Example 5.10, form an ascending geometric chain in this enumeration. Now, while the pythagorean form (1) leads to the familiar fifth-oriented scales, form (2) is very unsuitable from the point of view of achieving as many perfect fifth intervals as possible; the two forms (3) are the scale building blocks of perfect diatonic scales, in which a balanced mixture of perfect thirds and perfect fifths is sought and achieved (for more details, see for example [16]).

In the following, we will deal primarily with the pythagorean tetrachords (1) – the interesting cases (3) differ rather only in 'finer inner structures'. However, these are undoubtedly of interest, namely when semitonal or even sub-semitonal analyses of, for example, gregorian melodic progressions take place or when the ambitus of various neumes is explored – keyword would be, for example, the prominent

Question *'Should the tone si-bemolle be sung in a melodic progression or si-naturale – and: what is that anyway?'*

For a concise explanation of formal processes, we label the three genera of pythagorean diatonic tetrachords as follows:

$$T_{\text{phry}} = \left[\begin{bmatrix} \frac{1}{2} & 1 & 1 \end{bmatrix} \right] = [243, 256] \oplus [8, 9] \oplus [8, 9],$$

and accordingly the lydian T_{ly} and the dorian pythagorean diatonic tetrachord T_{do} are described. A single whole tone (8 : 9), which takes on the functional role of an interval connecting the tetrachords appears in the symbol ①.

About the interplay of these tetrachords there are the following useful aids, the correctness of which one quickly recognizes. *(Note again that the 'inverse' of a proportion chain lists the magnitudes in the reverse order – not the steps).*

Proposition 5.2 (Structural Rules for Pythagorean Diatonic Tetrachords)

For the processes of the chain-compositions, the processes of reciprocal and inverse chains of ancient Greek tetrachords we can easily find these formulas

1. The composition of whole tones and Pythagorean diatonic tetrachords has some useful structural laws, such as these

$$1 \oplus \left[\left[\tfrac{1}{2} \, 1 \, 1\right]\right] = \left[\left[1 \, \tfrac{1}{2} \, 1\right]\right] \oplus 1 \Leftrightarrow 1 \oplus T_{phry} = T_{do} \oplus 1,$$

$$1 \oplus \left[\left[1 \, \tfrac{1}{2} \, 1\right]\right] = \left[\left[1 \, 1 \, \tfrac{1}{2}\right]\right] \oplus 1 \Leftrightarrow 1 \oplus T_{do} = T_{ly} \oplus 1,$$

2. For the process of reciprocals we have the formulas:

$$\left[\left[\tfrac{1}{2} \, 1 \, 1\right]\right]^{rez} \cong \left[\left[1 \, 1 \, \tfrac{1}{2}\right]\right] \Leftrightarrow (T_{phry})^{rez} \cong T_{ly},$$

$$\left[\left[1 \, 1 \, \tfrac{1}{2}\right]\right]^{rez} \cong \left[\left[\tfrac{1}{2} \, 1 \, 1\right]\right] \Leftrightarrow (T_{ly})^{rez} \cong T_{phry},$$

$$\left[\left[1 \, \tfrac{1}{2} \, 1\right]\right]^{rez} \cong \left[\left[1 \, \tfrac{1}{2} \, 1\right]\right] \Leftrightarrow (T_{do})^{rez} \cong T_{do},$$

3. The process of inversion, on the other hand, only leads to the same tetrachord constructed in downward order - that is, the steps are also inverted, but in the opposite direction, so that we have the formulas

$$(T_{phry})^{inv} = \left[\left[\tfrac{1}{2} \, 1 \, 1\right]\right]^{inv} = (9 : 8) \oplus (9 : 8) \oplus (256 : 243),$$

$$(T_{ly})^{inv} = \left[\left[1 \, 1 \, \tfrac{1}{2}\right]\right]^{inv} = (256 : 243) \oplus (9 : 8) \oplus (9 : 8),$$

$$(T_{do})^{inv} = \left[\left[1 \, \tfrac{1}{2} \, 1\right]\right]^{inv} = (9 : 8) \oplus (256 : 243) \oplus (9 : 8).$$

In the following the term **'octochords'** comes up repeatedly. According to the word, such an octochord is merely an 8-stringed lute – an eight-fold monochord. We specify this in general to the situation of an octave: the general octochord is a heptatonic octave scale of eight tones in seven different steps. To this is added a condition of internal construction.

This and the **systema teleion** of ancient greek music theory are thus determined by the following definition:

Definition 5.3 (The Octochord and the Systema Teleion)

An **octochord** is a heptatonic octave scale – thus a sequence of 7 step intervals respectively proportions, structured in the form of an composition of exactly two equal tetrachords (dorian, phrygian or lydian) and one – necessarily – pythagorean whole tone 8 : 9.

The sequence of tones or intervals – usually notated downwards – or also the proportion chain **'systema teleion'** is the arrangement of four phrygian downward tetrachords and two downward whole-tone steps according to the following pattern:

Consequently, the range of the systema teleion is exactly two octaves and it consists of the composition of two identical octave scales **(octochords)**, which – written as proportion chains – have the structure in the upward direction:

Octochord of the systema teleion

$$[8, 9] \oplus T_{\text{phry}} \oplus T_{\text{phry}}.$$

According to Theorem 2.2 – the octave principle – all tone pairs of the systema teleion that differ by 7 steps are just octaves 1 : 2 respectively 2 : 1.

This closer characterization of the octochord already shows that a functional structure of the scales is intended. For the knowledge of the gregorian modes this observation is of great use.

The ancient description of this system, which is regarded as a universal tonal stock, also provides excellent information about this – even if it takes some effort to understand the large variety of all tone and step designations and their meaning. The diagram of Fig. 5.3 shows this descending double-octave system 'systema teleion' with the corresponding tone-step signs of the vocal and instrumental notation, thus giving an impression of the stringency of ancient music theory.

▶   The ancient greek octave scales and the related ecclesiastical modes now developed as octochords – in that a heptatonic selection (i.e., of the range of an octave) was taken from what was considered to be a complete construction kit of musical tones and tone steps.

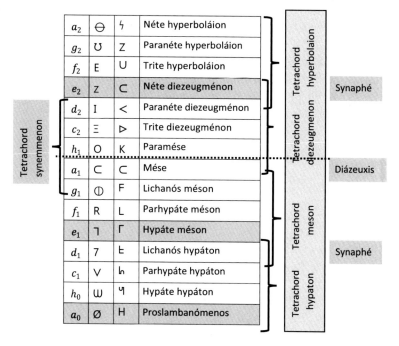

Fig. 5.3 The Greek systema maximum (systema teleion)

The manner in which and under what rules this extraction had to take place actually deserves its own, more expansive, consideration. However, if we approach the matter from the result, we nevertheless find more streamlined approaches to the ancient greek and later ecclesiastical tonal scales (modes), and we want to outline these now.

Depending on the approach and interpretation, there are 8, 9 or 12 different 'modes' and their constructions use the following rules:

Definition 5.4 (Rules for the Basic Architecture of Ecclesiastical Modes)
1. The amount of sound is determined by the (upward notated) proportion chain of the system

$$\left([8,9] \oplus T_{\text{phry}} \oplus T_{\text{phry}}\right) \oplus \left([8,9] \oplus T_{\text{phry}} \oplus T_{\text{phry}}\right)$$

This is the composition of two octochords.
2. An ancient greek octave scale consists – as an octochord – of the composition of two equal tetrachords and a whole tone 8 : 9 – the pythagorean tonos. The laws of affixation of Theorem 5.2 are effective here.
3. These two tetrachords are of **diatonic** gender (T_{phry}, T_{do} or T_{ly}).

There are basically three different approaches to deriving the octochords, the ancient greek scales and the gregorian scales, from the Systema teleion. We will outline these methods:

(A) **The combination method:** to two equal tetrachords (T_{phry}, T_{do} or T_{ly}) add a tonos 8 : 9, what we want to note by the number 1 - sometimes also by the symbol ①. This can be done at the three positions: at the beginning (1-tt), in the middle (t-1-t) and at the end (tt-1). Therefore we have with these signatures the octochords

$$\text{Oct}^{1-tt}_{\text{phry}} = 1 \oplus \left[\left[\tfrac{1}{2}\ 1\ 1\right]\right] \oplus \left[\left[\tfrac{1}{2}\ 1\ 1\right]\right],$$

$$\text{Oct}^{t-1-t}_{\text{phry}} = \left[\left[\tfrac{1}{2}\ 1\ 1\right]\right] \oplus 1 \oplus \left[\left[\tfrac{1}{2}\ 1\ 1\right]\right],$$

$$\text{Oct}^{tt-1}_{\text{phry}} = \left[\left[\tfrac{1}{2}\ 1\ 1\right]\right] \oplus \left[\left[\tfrac{1}{2}\ 1\ 1\right]\right] \oplus 1,$$

$$\text{Oct}^{1-tt}_{\text{do}} = 1 \oplus \left[\left[1\ \tfrac{1}{2}\ 1\right]\right] \oplus \left[\left[1\ \tfrac{1}{2}\ 1\right]\right],$$

$$\text{Oct}^{t-1-t}_{\text{do}} = \left[\left[1\ \tfrac{1}{2}\ 1\right]\right] \oplus 1 \oplus \left[\left[1\ \tfrac{1}{2}\ 1\right]\right],$$

$$\text{Oct}^{tt-1}_{\text{do}} = \left[\left[1\ \tfrac{1}{2}\ 1\right]\right] \oplus \left[\left[1\ \tfrac{1}{2}\ 1\right]\right] \oplus 1,$$

$$\text{Oct}^{1-tt}_{\text{ly}} = 1 \oplus \left[\left[1\ 1\ \tfrac{1}{2}\right]\right] \oplus \left[\left[1\ 1\ \tfrac{1}{2}\right]\right],$$

$$\text{Oct}^{t-1-t}_{\text{ly}} = \left[\left[1\ 1\ \tfrac{1}{2}\right]\right] \oplus 1 \oplus \left[\left[1\ 1\ \tfrac{1}{2}\right]\right],$$

$$\text{Oct}^{tt-1}_{\text{ly}} = \left[\left[1\ 1\ \tfrac{1}{2}\right]\right] \oplus \left[\left[1\ 1\ \tfrac{1}{2}\right]\right] \oplus 1.$$

This results formally in 9 octochords - but there are exactly seven different scales - regarded as an sequence of the step numbers 1 and 1/2 respectively of the heptatonic architecture build by the six whole tones and the two halftones. Task: Find the two pairs of Octochords which leads to the same scales notation.

However, there are still musical distinctions between these octochords which are considered to be similar from the superficial point of view of the architecturally determined whole tone and halftone structure. The Gregorian key-alphabet – the Gregorian modes of the Octoechos – is then obtained under the aspect of an additional musical-functional meaning of certain tone steps. Table 5.4 now shows these possible realisations of the octochord scales, whereby we take the "white keys" (c - d - e - f - g - a - h - c) of the keyboard as the basis of the tone model space, and we would like to give a brief explanation of this table. The 4. column indicates the "Modus Gregorianus" – i.e. the church scales. It is traditionally notated by the Roman numeral sign and you will allways find it notated at the beginning of the chants of the Gregorian editions. The knowlegeable chorister can thus recognise where the structural tones (finalis, recitativ tone (Tenor) and some more) are to be found. This in turn determines the way in which the chorale is sung - but also the combination of different parts of the choral complex (Antiphon, Hymn Psamody). In the

Table 5.4 Construction of the octochords and Gregorian modes according to the combination method

Tetrachords used	Octochord = 2 Tetrachords + 1 Wholetone	Model scale	Modus gregor.	Greek scale
Phrygian	$① \oplus [\![\frac{1}{2}\, 1\, 1]\!] \oplus [\![\frac{1}{2}\, 1\, 1]\!]$	a-h-c-d-**e**-f-g-a	IX	Aeolian
Phrygian	$[\![\frac{1}{2}\, 1\, 1]\!] \oplus ① \oplus [\![\frac{1}{2}\, 1\, 1]\!]$	e-f-g-a-**h**-**c**-d-e ē-f-g-**a**-h-c-d-e	III X	Phrygian Hypo-aeolian
Phrygian	$[\![\frac{1}{2}\, 1\, 1]\!] \oplus [\![\frac{1}{2}\, 1\, 1]\!] \oplus ①$	h-c-d-**e**-f-g-**a**-h h-c-d-ē-f-g-a-h	IV --	Hypo-phrygian Locrian
Lydian	$① \oplus [\![1\, 1\, \frac{1}{2}]\!] \oplus [\![1\, 1\, \frac{1}{2}]\!]$	**f**-g-a-h-**c**-d-e-f f̄-g-a-**h**-c-d-e-f	V --	Lydian Hypo-locrian
Lydian	$[\![1\, 1\, \frac{1}{2}]\!] \oplus ① \oplus [\![1\, 1\, \frac{1}{2}]\!]$	c-d-e-**f**-**g**-a-h-c c-d-e-f̄-**g**-a-h-c	VI XI	Hypo-lydian Ionian
Lydian	$[\![1\, 1\, \frac{1}{2}]\!] \oplus [\![1\, 1\, \frac{1}{2}]\!] \oplus ①$	g-a-h-**c**-d-e-f-g	XII*	Hypo-ionian
Dorian	$① \oplus [\![1\, \frac{1}{2}\, 1]\!] \oplus [\![1\, \frac{1}{2}\, 1]\!]$	g-a-h-c-**d**-e-f-g	VII	Mixolydian
Dorian	$[\![1\, \frac{1}{2}\, 1]\!] \oplus ① \oplus [\![1\, \frac{1}{2}\, 1]\!]$	**d**-e-f-g-**a**-h-c-d d̄-e-f-**g**-a-h-**c**-d	I VIII	Dorian Hypo-mixolydian
Dorian	$[\![1\, \frac{1}{2}\, 1]\!] \oplus [\![1\, \frac{1}{2}\, 1]\!] \oplus ①$	a-h-c-**d**-e-**f**-g-a	II	Hypo-dorian

Table the Finalis is double underlined and the Tenor is formed in bold and italic type. The result of all these possible attachments is now given in Table 5.4:

(B) **The octochord method:** Beginning with the lowest note of the systema teleion (the ***proslambanómenos*** (a)), one builds an octochord in such a way that its set of notes belongs to the systema teleion and precisely for this, one must apply the equations of the preceding Theorem 5.2. This is done successively on all seven pitches. It results in the same 9 models as in method (A).

The result is shown in Table 5.5.

To discuss two examples: If we start at the pitch $h = $ si, the systema teleion already determines the sequence of subsequent steps – in this case, the pattern

$$\frac{1}{2} - 1 - 1 - \frac{1}{2} - 1 - 1 - 1.$$

Obviously, the octochord starts the sequence with T_{phry} and then this tetrachord must immediately follow; the final whole tone then leads to the tonic octave (si). If we start on the tonic $g = $ sol, the sequence of the systema teleion

Table 5.5 Construction of the octochords according to the octochord method

Tonic	Octochord(s)	Step-diagram	Ancient Greek
a (la)	$\textcircled{1} \oplus T_{\text{phry}} \oplus T_{\text{phry}}$ $T_{\text{do}} \oplus T_{\text{do}} \oplus \textcircled{1}$	$1 - \frac{1}{2} - 1 - 1 - \frac{1}{2} - 1 - 1$	Aeolian/Hypodorian
h (si)	$T_{\text{phry}} \oplus T_{\text{phry}} \oplus \textcircled{1}$	$\frac{1}{2} - 1 - 1 - \frac{1}{2} - 1 - 1 - 1$	Locrian/hypophrygian
c (do)	$T_{\text{ly}} \oplus \textcircled{1} \oplus T_{\text{ly}}$	$1 - 1 - \frac{1}{2} - 1 - 1 - 1 - \frac{1}{2}$	Ionian/hypolydian
d (re)	$T_{\text{do}} \oplus \textcircled{1} \oplus T_{\text{do}}$	$1 - \frac{1}{2} - 1 - 1 - 1 - \frac{1}{2} - 1$	Dorian/hypomixolydian
e (mi)	$T_{\text{phry}} \oplus \textcircled{1} \oplus T_{\text{phry}}$	$\frac{1}{2} - 1 - 1 - 1 - \frac{1}{2} - 1 - 1$	Phrygian/Hypoaeolian
f (fa)	$\textcircled{1} \oplus T_{\text{ly}} \oplus T_{\text{ly}}$	$1 - 1 - 1 - \frac{1}{2} - 1 - 1 - \frac{1}{2}$	Lydian/Hypolocryan
g (sol)	$\textcircled{1} \oplus T_{\text{do}} \oplus T_{\text{do}}$ $T_{\text{ly}} \oplus T_{\text{ly}} \oplus \textcircled{1}$	$1 - 1 - \frac{1}{2} - 1 - 1 - \frac{1}{2} - 1$	Mixolydian/hypoionian

$$1 - 1 - \frac{1}{2} - 1 - 1 - \frac{1}{2} - 1$$

for lining an octochord structure is present.

Case A: If we interpret the first whole tone step as a connecting whole tone $\textcircled{1}$, then it is followed by two dorian tetrachords T_{do}.

Case B: Otherwise, the start is the lydian tetrachord T_{ly}. The rest chain $1 - 1 - \frac{1}{2} - 1$ could be of the form $\textcircled{1} \oplus T_{\text{do}}$ or of the form $T_{\text{ly}} \oplus \textcircled{1}$; the former can be ignored since there must be two equal tetrachords in the octochord. Thus we have two solutions of a structuring of the scale progression.

(C) **The step method:** one runs a combinatorial game of a scale construction of (2) half steps and (5) whole steps under the following rules:
  1. No two semitones may occur in succession.
  2. No more than three whole tone steps may occur in succession.
  3. In the octave balance, there are exactly 5 whole steps and 2 half steps.

Rules 1 and 2 also include possible scale continuations (octave transpositions) upwards or downwards.

We get a scale family of exactly 7 different models, shown in Table 5.6, where the last (3rd) condition is actually even obsolete – the octave balance requires exactly this constellation (see for example [16]).

One simply starts by systematically writing down all possible step series from the stock of whole and half steps, and then, due to the rules, you can exclude all cases except the seven noted down.

Table 5.6 Construction of the ancient Greek scales according to the step method

Step by step	Scale	Octochord	Ancient Greek keys
$1-1-1-\frac{1}{2}-1-1-\frac{1}{2}$	f – f'	Okt_{ly}^{1-tt}	Lydian or Hypolocrian
$1-1-\frac{1}{2}-1-1-\frac{1}{2}-1$	g – g'	Okt_{do}^{1-tt} respectively Okt_{ly}^{tt-1}	Mixolydian or Hypoionian
$1-1-\frac{1}{2}-1-1-1-\frac{1}{2}$	c – c'	$\mathrm{Okt}_{ly}^{t-1-t}$	Ionian or Hypolydian
$1-\frac{1}{2}-1-1-\frac{1}{2}-1-1$	a – a'	$\mathrm{Okt}_{phry}^{1-tt}$ respectively Okt_{do}^{tt-1}	Aeolian or Hypodorian
$1-\frac{1}{2}-1-1-1-\frac{1}{2}-1$	d – d'	$\mathrm{Okt}_{do}^{t-1-t}$	Dorian or Hypomixolydian
$\frac{1}{2}-1-1-1-\frac{1}{2}-1-1$	e – e'	$\mathrm{Okt}_{phry}^{t-1-t}$	Phrygian or Hypoaeolian
$\frac{1}{2}-1-1-\frac{1}{2}-1-1-1$	h – h'	$\mathrm{Okt}_{phry}^{tt-1}$	Locrian or Hypophrygian

The three methods ultimately lead to a common overall picture. In the case of the **ancient greek tone scales,** all doublings apparently emerge through the syllable 'hypo'. There is a system behind this, which will be described in the following theorem.

The **gregorian modes** are determined first and foremost by the position of the so-called **finalis** (printed in bold and underlined in Table 5.4) and by the position of the **tenor** (recitative tone, co-finalis; in bold italics in Table 5.4). A description given in the gregorian textbooks corresponds to something like this:

▶ **'Authentic'** are the **odd** modes – with the finalis low and the tenor a fifth above. The **even** modes are called **'plagalic'** – in them the finalis is approximately in the middle of the melody and the tenor is a third or fourth above it. Another classification of gregorian modes – based on the so-called Octoechos – makes use of the basic types of the finalis-step environment

Protus, Deuterus, Tritus and **Tetrardus**

authentic or plagal in each case – in order that – via these Gregorian 'earliest modes' to get to the eight main modes (see [11, p. 70 ff.]).

When speaking of symmetries and the same for 'keys' in what follows, this refers to the step-proportion chains that represent them. We summarize our foregoing thoughts in the following theorem:

Theorem 5.2 (The Systema Teleion and the Ecclesiastical Keys (Modes))
Assuming that we are dealing with a unified whole tone, which is why the **pythago-rean diatonic tetrachord** then defines the **systema teleion**, the following statements hold:

1. **Methodology:** By means of the octochord method as well as the combination method, the **ancient greek scales** as well as the **gregorian modes** can be determined. The step method proves that there cannot be further scale structures of the systema teleion.
2. **Balances:** The following totals result in the systema teleion:
    1. There are exactly 7 different step sequences.
    2. There are exactly 9 different octochords.
    3. There are exactly 12 gregorian modes.
    4. There are exactly 14 ancient greek modes.
3. **Authentic-Plagal:** One goes from an authentic to its plagalic gregorian form by means of the following rule:
    Authentic-Plagal Rule: If the octochord of the authentic key (I, III, V or VII) is present in its structure, the corresponding plagal form starts with the beginning of the second tetrachord of this octochord.
4. **Modes and their hypo-modes:** One gets from an ancient greek mode (aeolian, locrian, ionian, dorian, phrygian, lydian or mixolydian) to its 'hypoform' by the same rule as in (3.).
5. **Octochord symmetries:** At the level of step proportions, the symmetry relations hold:

$$1.\ \mathrm{Oct}_{\mathrm{do}}^{t-1-t} \cong \left(\mathrm{Oct}_{\mathrm{do}}^{t-1-t}\right)^{\mathrm{rez}}$$

$$2.\ \mathrm{Oct}_{\mathrm{ly}}^{1-tt} \cong \left(\mathrm{Oct}_{\mathrm{phry}}^{tt-1}\right)^{\mathrm{rez}} \text{ and also } \mathrm{Oct}_{\mathrm{ly}}^{t-1-t} \cong \left(\mathrm{Oct}_{\mathrm{phry}}^{t-1-t}\right)^{\mathrm{rez}}$$

$$3.\ \mathrm{Oct}_{\mathrm{do}}^{1-tt} \cong \mathrm{Oct}_{\mathrm{do}}^{tt-1} \cong \left(\mathrm{Oct}_{\mathrm{phry}}^{1-tt}\right)^{\mathrm{rez}} \cong \left(\mathrm{Oct}_{\mathrm{ly}}^{tt-1}\right)^{\mathrm{rez}}.$$

This means that the dorian mode is symmetrical: the step proportion succession upwards is the same as in the downward movement. No other key has this symmetry. For the other octochords, we have the symmetries listet in the formulas 2. and 3.

From these, one can point out some hidden relations – both in the symbolic combinatorics and in the scale structures.

The proof of this theorem consists of an observational compilation of the three Tables 5.4, 5.5, and 5.6, which – obeying the three methods – bring together scales and octochords as well as step geometries.

We would also like to highlight that all these results apply first of all exclusively to the very simplest model case of a bitonal systema teleion (tonos and limma). If, nevertheless, there is a difference between the whole tones (for example the important diatonic case that both whole tones 8 : 9 and 9 : 10 are present as well as that at the end still different semitonia inevitably occur), then some of our statements above can only be used in a modified way. More detailed investigations, however, can certainly be designed as smaller in-depth projects – using the shown methods.

Note: Si-naturale or Si-bemolle?

Finally – as well as following on from this – we illuminate a recurring theme in gregorian studies:

Question: Si-naturale or Si-bemolle – what is it all about?

Well, in the model of a perfect diatonic heptatonic scale, the notes sung in the gregorian melodies could be modeled – i.e. reproduced – on the white keys of a keyboard. The only exception is the tone si (i.e. 'h'):

Problem: Here we find both (!) alternatives of a whole tone on a – hence even leading to **two** different tones h of the white key as well as simultaneously define several 'half tones' on a, which then lead to 'b' – and gregorianically one calls (actually all) these half tones 'si-bemolle'.

We now need to know that the earlier notation was not actually a notation: only a paleographic neume script described and describes in a rather coded way which tone steps from tone to tone would probably be meant. This can be recognized in our Fig. 5.4 of the "Rorate introitus", in which the 'neumes' consist of the mysterious signs above and below the later established stave system. As far as the question of the two forms of the tone 's'' is concerned, there is indeed everywhere and always the discussion – fueled by comparative manuscripts of the neumes – of what would probably be right and what would probably be wrong.

Our Fig. 5.4 shows the famous introit 'Rorate' of the fourth Sunday of Advent in the vatican notation, in which a si-bemolle is notated at the beginning after the fifth-pes 're-la' – but it could also be a si-naturale. The first version sounds like 'D minor', the second like 'dorian', tonus I of the ecclesiastical mode.

Fig. 5.4 The beginning of the "Rorate Introitus" (Source: Graduale Triplex)

The gaze that guides these discussions, however, is – consciously or unconsciously – devoted to the familiar heptatonic scale structure. Therefore, a sung rendition always takes place in the bivalent alternative 'semitone' or 'whole tone'. There are a number of critical questions and gregorian research is very intensively devoted to **melodic restitution.**

1. **Problem:** Which gradation is meant when we think of the sequence of tones?

$$a \rightarrow si - b - molle\ (b) \rightarrow si - naturale\ (h)$$

It is clear that these two intervals cannot be of equal size; the division of the tonic 8 : 9 into two equal halves leads to the proportion chain

$$8 : z_{geom}(8,9) : 9 \text{ with the irrational number } z_{geom}(8,9) = 6\sqrt{2} \approx 8,485\ldots,$$

and considering the then prevailing pythagorean interval theory, one would rather think of the extremely asymmetrical apotome-limma division.

2. **Problem:** It could also be that the ancient greek tetrachordics was nevertheless on the agenda and could give rise to interpretations with strange small microtonal gradations to consider the 'si-bemolle' as a minimal weakening or 'clouding' of the si-naturale – the voice just gives in a little – or it rises from *a* only 'imperceptibly'. We have already become acquainted with some strange microtonalities in tetrachordics (Sect. 5.5)...

3. **Problem:** Around this topic we also encounter considerations that possibly other melodic progressions could be interpreted from the neume writing: For example, a different melodic character would develop if in the word 'rorate' a higher 'do' – pronounced *c* – were placed instead of the tone 'si-bemolle' or 'si-naturale'. Indeed, this melodic formula is frequently found in late medieval church books!

In short: this and a number of other reasons we can identify in the partly bizarre chains of proportions, clearly give reason to consider the si-bemolle question under a considerably broader framework, which would undoubtedly permit far more and fundamentally different solutions than just the bivalent variant of the piano keyboard.

5.7   The Organ and the Calculus with Its Stop Proportions

With hardly any other instrument can we experience the relation between proportions and sounding music more directly than with the organ – let it be small or large symphonic instruments. Proportions can be seen in the geometry of the pipework as well as in the description of the entirety of the stops – the so-called **disposition.** If you take a look at the information about the stops of an organ, you will discover terms like

$$\text{Subbass } 16' - \text{or Principal } 8' - \text{or Octave } 4' - \text{or Fifth } 2\frac{2'}{3} - \text{or Third } 1\frac{3'}{5}$$

to name just the most common. What is behind it and what does the little apostrophe mean?

Well, these codes are called **'footnumbers'** to begin with, and the little signing apostrophe indicates the 'unit' of the number as the so-called 'footnumber' of the corresponding stops.

The organists then also say 'trumpet eight foot' and each knowledgeable knows that this is a stop whose pipes sound 'like a trumpet' and whose pitch frequencies correspond (to a large extent) to those of corresponding (equal-pitched) notes on the piano.

The same key – played with a 4-foot stops – would sound an octave higher and a note played with a 16-foot stop would sound an octave lower.

The concept of the **footnumber** is historical: it is the indication in the old unit of length 'foot' of how long the pipes would be at the lowest manual note – usually the great C. *For example, the pipe length of an (open) 8-foot stop for 'great C' measures just 8 feet.* The pipe lengths of a single register – that is, an 'instrument' contained within the organ – decrease as the pitch increases: physics says:

> If the length of the pipe is halved, the frequency of the tone is doubled – which is the same thing as saying that the tone sounds 1 octave higher than that of the 8-foot pipe.

Four octaves higher than the great *C,* the pipe of the 8 foot stop then only has the length of half a foot, i.e. a few centimeters. Hence we are dealing here with the 'length proportion'. The relation between pitch (frequency) and pipe length is a law of proportions, which of course obeys the monochord rules in its entirety. Adapted to the organ, we will describe as follows:

Theorem 5.3 (The Foot Number Rule of the Organ: 'The Organ Stops Mathematics')

For two pipes of the same stop (of which we assume idealistically that they are of similar geometrical construction – such as open at the top, same material and so on) there is the law of proportions between the pitches T_1 and T_2 (these are the fundamental frequencies) on the one hand and their lengths L_1 and L_2 on the other.

   Pitch proportion law of the organ:

$$L_1 : L_2 \cong T_2 : T_1$$

Conclusion: Given are any two organ stops, which we will named by

---

(continued)

Theorem 5.3 (continued)
   "Flute A" (with the foot number a – given in fractional arithmetic form),
   "Flute B" (with the foot number b – given in fractional arithmetic form),

then for each fixed tone key between its pitches T_a and T_b (measured in hertz, for example) there is the constant (key-independent) foot-number proportion that we write down in the following rule:
   General foot number rule of the organ:

$$T_a : T_b \cong b : a,$$

and we can also formulate this proportion equation fractionally in such a way:

$$T_a * a = T_b * b = \text{constant},$$

in other words: The product of pitch (T_a) **and footnote** (a) is – for each fixed key – the same regardless of the selected stops.
   Special case: The following applies specifically to the pitch relationship of an 'organ stop A' with the number of feet a to an 8-foot reference stop: When any key is struck, the following proportion applies to the pitches
   8-foot rule of the organ:

$$T_8 : T_a \cong a : 8,$$

from which one then usually determines the differing interval in relation to the 8-foot position. From this we obtain the tonal classification of an organ stop – concerning the comparative pitch.

   These statements ultimately follow from the well-known relation between lengths and pitches – just as we discussed with the monochord in Sect. 5.1. It makes no difference whatsoever whether it is a string or a sounding pipe – at least not in the physically guided (idealised) model.

▶   **Important**
   We would also like to note that the advantage of the proportion equation

$$T_a : T_b \cong b : a$$

is precisely that it applies **independently of the pitch** – that is, key-independently – while the transformation to the product equation

$$T_a * a = T_b * b$$

has the same value per key for any stop (i.e. it is **independent of the stop**) – but is proportional to the pitch of the keyboard per stop.

Let's leave it like that and direct our attention to practical examples.

If we use these basic rules to study the tonal proportions of an organ on the basis of its foot-number characteristics, it is very helpful – indeed necessary – to have the **fractional arithmetic** data of the perfect intervals at hand – at least major thirds (4 : 5 or in frequency measure 5/4) and fifths (2 : 3 or 3/2) as well as, first and foremost, the most important and omnipresent octave markings (16 – 8 – 4 – 2 – 1) should be quickly at hand.

Example 5.21 Stop Proportions of the Organ

1. The stop woodland flute 2′ is two octaves (i.e. a double octave) above the note for each key if an 8-foot stop were played on it:

$$T_8 : T_2 \cong 2 : 8 \cong 1 : 4.$$

2. The stop fifth $2\,\tfrac{2}{3}'$ shows the following interval position

$$T_8 : T_{2\frac{2}{3}} \cong 2\frac{2}{3} : 8 = \frac{8}{3} : 8 \cong 1 : 3 \cong (2 : 6) \cong (2 : 3) \odot (1 : 2)$$

and is thus 1 octave and 1 perfect fifth (which is a duodecimal) above the 8 foot note. It is also immediately clear that:

$$T_4 : T_{2\frac{2}{3}} \cong 2\frac{2}{3} : 4 \cong \frac{8}{3} : 4 \cong 8 : 12 = 2 : 3,$$

and that's a fifth above the 4-foot stop.

3. For example, for the third $1\,\tfrac{3}{5}'$ we have the following comparative proportions:

$$T_2 : T_{1\frac{3}{5}} \cong 1\frac{3}{5} : 2 = \frac{8}{5} : 2 \cong 8 : 10 \cong 4 : 5 \text{ (comparison with the 2-foot stop)},$$

$$T_8 : T_{1\frac{3}{5}} \cong 1\frac{3}{5} : 8 = \frac{8}{5} : 8 \cong 8 : 40 \cong 1 : 5 \text{ (comparison with the 8-foot stop)},$$

$$T_{2\frac{2}{3}} : T_{1\frac{3}{5}} \cong 1\frac{3}{5} : 2\frac{2}{3} = \frac{8}{5} : \frac{8}{3} \cong 3 : 5 \left(\text{comparison with the } 2\frac{2}{3}\text{-foot stop}\right).$$

4. We encounter completely unusual foot-number ratios in the disposition of **italian organs**. Many of them contain a collection of extremely high-pitched pipes in the so-called 'ripieno'.

In the **Chiesa Parrocchiale di San Martino Vescovo** in the small town of Sarnico, for example, we find an organ by Giovanni Giudici from Bergamo, which in its manual "Hauptwerk" – among many other exotics – also contains the stop

$$\text{"Due di Ripieno"} \frac{1'}{3} - \frac{1'}{4}$$

It is a small **mixture form,** consisting of two pipes per key – one at 1/3 foot, the other at 1/4 foot. Which intervals, pitches are given, compared to the 8 – or to the 1-foot position?

Firstly, 1/4-foot: This is obviously 5 octaves above the 8-foot, as we can see from the halving order $8 - 4 - 2 - 1 - 1/2 - 1/4$ as well as according to our foot number rule, since it is

$$(1/4) : 8 \cong 1 : 32 = 1 : 2^5.$$

Secondly, 1/3-foot: Here we make the comparison to the 1/2-foot position, which is already 4 octaves above the 8-foot position and thus in the 'vicinity' of the stop in question. According to the foot number rule

$$(1/3) : (1/2) \cong 2 : 3,$$

and therefore this pipe sounds 1 fifth above the fourth octave above the 8-foot position.

Together, 1/3- and 1/4-foot hence form a remote fourth (namely 5 octaves up with a fourth below), an adventurously extreme position. Even the middle note a – played on the 440 Hz pitched octave in the 8-foot position – sounds 5 octaves higher in the ¼-foot: that is

$$440 * 2^5 = 440 * 32 = 14080 \ \text{Hz} \ \ (\text{Hertz}),$$

which is almost too high, even for dogs – not to mention keyboard tones of even higher octaves.

5. **The Great Third in Schwerin Cathedral:** A diametrically opposed situation to example (4) is found in the famous **Ladegast organ of Schwerin Cathedral.** There, in the pedal division, there is a just tempered great third which, together with the 16-foot basses, results in a so-called **acoustic 64-foot,** that corresponds to tones whose vibrational frequencies would have to call the building architects to the scene. Assuming that the designation of the stops of this great third had been lost and that the organ builder only knew about the circumstance of this phenomenon, then the following question would arise

Question: What must the foot specification be, so that the pairing of a 16-foot stop with this 'great third' results in the acoustic 64-foot?

Answer: First, the third must be a **perfect or just** major third over a 16-foot stop tone. We'll clarify later why the overlays then create the famous 64-foot. Again, we can use the foot-number formula. Accordingly, the proportion equation must hold:

$$T_{16} : T_a \cong a : 16 \cong 4 : 5.$$

From this follows immediately for the foot number a

$$a : 1 \cong (16 * 4) : 5 \cong \frac{64}{5} : 1 \cong \frac{128}{10} : 1 \Leftrightarrow a = \frac{128}{10} = 12\frac{4}{5},$$

and the problem is solved.

The wealth of so-called **'aliquots'** on great organs – i.e. all those stops whose notes do not differ by **octaves** (i.e. whose foot numbers do not have the form 2^n with $n = 0, \pm 1 \pm 2$) – is sometimes considerable. Aliquots not only provide acoustic overlay effects – as seen in example (5) – they are also instrumental in timbre shaping (namely by influencing the overtone structure – as in example (4)). Here is another demonstration:

6. The new gigantic organ of the most important portuguese church in **Fatima** has no less than five manuals, pedal and a total of about 90 stops (cf. Ars Organi 2/2017, [21]). In the swell (Manual III) there is a five-fold mixtures stop **(terziana)**, whose composition has a set of aliquots with the following exotic-seeming foot numbers:

$$5\frac{1}{3} - 3\frac{1}{5} - 2\frac{2}{3} - 1\frac{1}{7} - \frac{8}{9}.$$

Each time a key is pressed, a battery of tones consisting of these five individual tones is sounded.

Question: What frequency multiples (overtones) above a 16-foot fundamental do these aliquots correspond to?

Answer: We use the foot number rule this time in the inverted form

$$T_a : T_{16} \cong 16 : a$$

and insert the aliquot footnumbers for a one after the other. Then for the overtone sequence T_a/T_{16} the value sequence

$$3 = \frac{16}{5\frac{1}{3}}, \ 5 = \frac{16}{3\frac{1}{5}}, \ 6 = \frac{16}{2\frac{2}{3}}, \ 14 = \frac{16}{1\frac{1}{7}}, \ 18 = \frac{16}{\frac{8}{9}}.$$

is obtained. This overtone series can be conveniently used for tonal description; we decompose these ratios into prime factors and obtain

$$3 - 5 - 6 - 14 - 18 \ \text{respectively} \ 3 - 5 - (2 * 3) - (2 * 7) - (2 * 3 * 3).$$

Subsequently the respective fundamental frequencies are now the $3 - 5 - 6 - 14 -$ or 18-fold of the played fundamental in 16-foot position. From the prime number factorization as well as from the fact that frequency measure products correspond to stratifications of intervals, the inversions of the proportions finally mean

$T_{16} : T_a \cong 1 : 3 \ \rightleftarrows$ fifth above the octave (duodecime, twelfth),
$T_{16} : T_a \cong 1 : 5 \ \rightleftarrows$ major third above the 2. octave,
$T_{16} : T_a \cong 1 : 6 \ \rightleftarrows$ fifth above the 2. octave,
$T_{16} : T_a \cong 1 : 14 \ \rightleftarrows$ just seventh above the 3. octave,
$T_{16} : T_a \cong 1 : 18 \ \rightleftarrows$ whole tone $(8 : 9)$ above the 4. octave.

With a single keystroke you have activated this overtone spectrum. By the way, the two sub-sequences of $5\frac{1}{3}$-foot and $3\frac{1}{5}$-foot already provide for an acoustic 16-foot-stop.   ◀

Conclusion From these proportions one can see very nicely the positions of the stops among each other: Example (3) shows that this 'third $1\frac{3}{5}$' sounds raised by a perfect diatonic major sixth $(3 : 5)$ above this 'fifth $2\frac{2}{3}$'. Interesting sound effects can be achieved in this way – such as the construction of a

"**Cornet**" – resulted from combinations with $8 -, 2\frac{2}{3} -$ and $1\frac{3}{5} -$ foot stops.

This corresponds to a spread major chord in just temperament. When the c_0 key is pressed, the following chord is played

$$c_0 - g_1 - e_2 \ \rightleftarrows \ \text{Duodecime} \oplus \text{major sixth},$$

thus a spread major chord. The foot-number and overtone relation is even more impressive in example (6) – on a single tone key c_0, the chord is created

$$c_0 - g_1 - e_2 - g_2 - b_3^* - d_4$$

⇌ Duodecime ⊕ major sixth ⊕ minor third ⊕ ecmelic minor third ⊕ ecmelic major third.

One can also interpret this chord as a combination of a spread major chord (here: $c_0 - g_1 - e_2 = C -$ major) with a nearly minor chord added on a high fifth ($g_2 - b_3^* - d_4 = g -$ minor); the note b_3^*, however, lies about halfway between a and b; in the fundamental octave, the proportion $4 : 7$ has the cent-measure of $968, 8$ ct; in effect, the overall chord is of the type of a (strongly) *diminished* C^7 -seventh chord with an added high ninth (d). Mind you, we are talking about a 'chord' that lies on a single key. Nevertheless, due to the appropriate choice of amplitude of these overtones, the ear perceives the overall sound as **unison** embedded in the timbre of a cornet. It doesn't sound bad if you know how to use it.

Remark: The Phenomenon of Acoustic Sub-Octave Magic on the Organ

In organ building, the following phenomenon is used in order to achieve a lower pitch – by one or two octaves – from two low-sounding stops, without a **separate stops** being installed for this purpose. This phenomenon can be observed like this:

1. If you play a 16-foot stop and a **justly tuned** fifth – i.e. a $10\frac{2}{3}$-foot stop – together, you get the impression that you are also playing a 32-foot register – without the need for a separate pipe.
2. If one also plays a 16-foot stop together with a just major third above it – which, according to our example, is a $12\frac{4}{5}$ -foot stop – the result is an undertone that corresponds to a 64-foot stop (!).

If we consider that such 32-foot pipes or even 64-foot pipes are true giants – and they would (and certainly do) only have room in cathedrals and on low-loaders – it quickly becomes clear that this trick of acoustically superimposing two relatively moderate stops to pretend a non-existent low-frequency giant is very well used in organ building.

Question: How does this phenomenon occur?
Answer: Well, there are two main ways we could explain this:
(a) from the proportions of the 'overtones' of an ideal tone,
(b) from the physics of the superposition of two fundamental oscillations.

However, we will only touch approximately on both of these considerations:

To (a): Let's assume we have a 32-footer. The first harmonic (which is an octave higher) is in proportion 1 : 2 to this – it corresponds to a 16-foot; the second has the proportion 1 : 3 and therefore corresponds to a $32/3 = 10\frac{2}{3}$ -foot (this is a perfect fifth above the octave, a duodecimal). It goes on like that. The proportion of the second to the first harmonic is therefore the perfect fifth interval 2 : 3.

Trick Now, if a 16-foot sounds together with the perfect $10\frac{2}{3}$-fifth above it, the ear believes to perceive the first and the second overtone (of a 32-foot) simultaneously – and it then 'simply hears the fundamental (32-foot) in addition'.

At the ratio 4 : 5 in the case of the great just third, we have a similar situation: 16-foot and $12\frac{4}{5}$-foot are the fourth and fifth harmonics of the 64-foot; the ear constructs the fundamental 64-foot backwards from these.

These explanations are only plausible considerations; however, they find their justification in the mathematical-physical consideration (b):

To (b): If two fundamental oscillations whose frequencies f_1 and f_2 are in the – exact (!) – ratio 2 : 3 are superposed, a new oscillation pattern is created which, in the case of this exact ratio (and only then), is periodic in the first instance. This period now has the frequency f_0 that can be derived from the simple formula

$$6f_0 = 2f_2 = 3f_1$$

The explanation is simple: between period of oscillation T and frequency f the reciprocal relation

$$T = 1/f$$

applies. Consequently for the oscillation period ratio the proportion-equations

$$T_1 : T_2 \cong 1/f_1 : 1/f_2 \cong 1/2 : 1/3 = 3 : 2 \Leftrightarrow 2T_1 = 3T_2,$$

will follow and the period of oscillation T_0 of the superposition is the smallest in which both fit integer. This is exactly expressed by the equation

$$T_0 = 2T_1 = 3T_2$$

and with $T_0 = 1/f_0$ we achieve the desired frequency relation.

However, it is also important to note that this period to the frequency f_0 must in practice hold over a certain time period – of at least a few seconds – otherwise the 2 : 3 ratio is immediately disturbed and a superposition cannot simulate a sub-octave. Therefore this fifth $10\frac{2}{3}$ must be tuned 'very' perfectly (justly) to the 16-foot position with, by the way, also clearly reduced amplitude, if one wants to achieve the tricky effect of the feigned 32-foot noticeably.

If the amplitude – i.e. the volume – is too big, one hears a disturbing fifth interval; if the fifth is not perfect to the 16-foot stop (because, for example, one keeps the intervals of the equal temperament instead of the just temperament), the desired effect diminishes noticeably.

32'-Trombone

$\text{♫} - \int e^{2\pi i \omega(t)}$

That's a science in itself – . but it always deals with proportions.

Appendix

Table of Proportions of the Most Important Rational (Perfect, Just) Intervals

The following Table A.1 lists the most common "pure" intervals (ordered by size): Here we give the proportions both in the proportion form used in the book and in the "frequency measure" form commonly used today. For the rest, we refer to the section "For use" in the introductory part of the book.

In the table we give the frequency measure b/a of a proportion $a : b$ equal in prime factorization (which here are mainly the prime numbers 2, 3, and 5). This allows a simple **tracing** of the defining representation of the interval as an **adjunctive composition** from

$$\text{Octaves } (1 : 2), \text{ Fifths } (2 : 3) \text{ and Thirds } (4 : 5).$$

Then the exponents of the prime factor 3 are exactly the number of required fifths and the exponent 5 the number of required thirds, and the number of octaves can be easily determined from this, because there is one downward octave in each proportion of the fifth and two in each proportion of the third; factors in the denominator ("negative exponents") correspond, of course, to subtractions of the corresponding intervals. An example may explain this in more detail:

Example

For Determining the Adjunctive Composition

In the **large chroma** $128 : 135$ we read the frequency measure $3^3 5^1 / 2^7$, and this is how we process it:

$$3^3 5^1 / 2^7 = 2^{-7} 3^3 5^1 = 2^{-7+3+2} * (3/2)^3 * (5/4)^1 = 2^{-2} * (3/2)^3 * (5/4)^1.$$

Then this corresponds to an interval which consists of the joining of three fifths $(2 : 3)$ and a third $(4 : 5)$ and the (subsequent) subtraction of two octaves. And one sees immediately that

© Springer-Verlag GmbH Germany, part of Springer Nature 2024
K. Schüffler, *Proportions and Their Music*,
https://doi.org/10.1007/978-3-662-65336-4

Table A.1 Table of the most important pure intervals and their proportions

Intervals	Proportion	Frequency measure (fraction)	Cent measure (ct)
Unison	1 : 1	$1/1 = 1$	0
Small chroma	24 : 25	$5^2/2^3 3^1$	70.5
Pythagorean limma	243 : 256	$2^8/3^5$	90.2
Large chroma	128 : 135	$3^3 5^1/2^7$	92.2
Diatonic semitone	15 : 16	$2^4/3^1 5^1$	111.7
Pythagorean apotome	2048 : 2187	$3^7/2^{11}$	113.7
Euler semitone	25 : 27	$3^3/5^2$	133.2
Diatonic whole tone	9 : 10	$2^1 5^1/3^2$	182.4
Pythagorean whole tone (Tonos)	8 : 9	$3^2/2^3$	203.9
Minor Pythagorean third	27 : 32	$2^5/3^3$	294.1
Perfect minor third	5 : 6	$2^1 3^1/5^1$	315.6
Perfect major third	4 : 5	$5^1/2^2$	386.3
Pythagorean major third (Ditonos)	64 : 81	$3^4/2^6$	407.8
Perfect (Pythagorean) fourth	3 : 4	$2^2/3^1$	498.0
Perfect (Pythagorean) fifth	2 : 3	$3^1/2^1$	702.0
Just minor sixth	5 : 8	$2^3/5^1$	813.7
Just major sixth	3 : 5	$5^1/3^1$	884.3
Nature (harmonic) seventh	4 : 7	$7^1/2^2$	968.8
Minor (diatonic) seventh	5 : 9	$3^2/5^1$	1017.6
Major (diatonic) seventh	8 : 15	$3^1 5^1/2^3$	1088.3
Octave	1 : 2	$2^1/1$	1200
Major (Pythagorean) ninth	4 : 9	$3^2/2^2$	1403.9
Major (diatonic) decime	2 : 5	$5^1/2^1$	1586.3
Double octave	1 : 4	$2^2/1$	2400

this balance is identical with the difference of the diatonic semitone 15 : 16 in the major whole tone 8 : 9 (this difference is in fact the original definition of the major chroma), since the simple equation applies

$$(9/8) * (15/16) = 135/128.$$

◄

In the Table A.1, the **cent measure** (rounded to one decimal place) is also given; this **logarithmic measure** is defined for all intervals - therefore also for all intervals with ancient proportion measure a: b (and consequently with frequency measure b/a) by the formula

Table A.2 Table of some commas of pure diatonics

Intervals	Proportion	Frequency measure (fraction)	Cent measure (ct)
Diaschisma	2025 : 2048	$2^{11}/3^4 5^2$	19.5
Syntonic comma	80 : 81	$3^4/2^4 5^1$	21.5
Pythagorean comma	524.288 : 531.441	$3^{12}/2^{19}$	23.5
Small diësis	125 : 128	$2^7/5^3$	41
Great diësis	625 : 648	$2^3 3^4/5^4$	62.5

$$I = [a, b] \text{ respectively } I \cong (a : b)$$

$$\Leftrightarrow \mathrm{ct}(I) = 1200 * \log_2(|I|) = 1200 * \log_2\left(\frac{b}{a}\right) = 1200 * \frac{\ln |I|}{\ln 2}$$

and here the symbol $\log_2$ stands for the "logarithm to base 2"; the symbol ln stands for the logarithm naturalis, and the formula can therefore also be used with simple pocket calculators to calculate the centvalues to the data a and b of a frequency measure fraction b/a.

In addition, especially in the theory of scales, which are formed by stratifications of

$$\text{Third } (4 : 5), \text{ Fifth } (2 : 3) \text{ and Octave } (1 : 2)$$

and which historically represent the center of temperament theory, the following characteristic **"commas"**-mostly in the quarter semitone range-are known and of great importance (see also [16]) (Table A.2):

Illustrations of the Analytical Mean Value Functions

In this part, the graphs of the four analytic functions appearing in the text are given in their respective relevant domains of definition; these are the following functions:

- the proportion function $y = f(x) = (x - a)/(b - x)$ (Fig. A.1),

- the inverse of the proportion function, the mean value-function
  $x = g(y) = a + \frac{y}{1+y}(b - a)$ (Fig. A.2),

- the hyperbola of Archytas $y = ab/x$ respectively $xy = ab$ (Fig. A.3),

- the harmonic hyperbola $y = ax/(2x - a)$ (Fig. A.4).

Fig. A.1 Progression of the
proportion function
$f(x) = (x - a)/(b - x)$

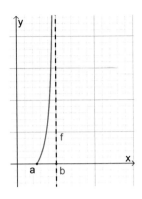

Fig. A.2 Progression of the
mean value function
$g(y) = a + \frac{y}{1+y}(b - a)$

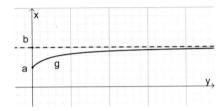

Fig. A.3 Progression of the
hyperbola of the archytas
$y = ab/x$

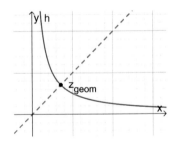

Fig. A.4 Progression of the
harmonic hyperbola
$y = ax/(2x - a)$

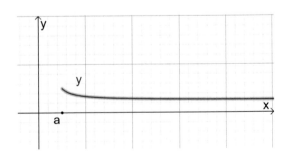

References

1. Ambros, A.W.: Geschichte der Musik. Leuckart, Leipzig (1911)
2. Assayag, G. (ed.): Mathematics and Music. Springer, Berlin (2002)
3. Aumann, G.: Euklids Erbe. Wiss. Buchgesellschaft, Darmstadt (2009)
4. Bühler, W.: Musikalische Skalen bei Naturwissenschaftlern der frühen Neuzeit. Lang, Frankfurt (2013)
5. Flotzinger, R.: Harmonie. Böhlau, Wien (2016)
6. von Freiherr Thimus, A.: Die harmonikale Symbolik des Altertums I. Georg Olms, Nachdruck, Hildesheim (1988). (Erstveröffentlichung 1868)
7. von Freiherr Thimus, A.: Die harmonikale Symbolik des Altertums II. Georg Olms, Nachdruck, Hildesheim (1988). (Erstveröffentlichung 1868)
8. Gericke, H.: Mathematik in Antike und Orient. Fourier, Wiebaden (1992)
9. Gow, J.: A Short History of Greek Mathematics. Chelsea, New York (1968)
10. Kayser, H.: Der hörende Mensch. Lambert Schneider, Berlin (1930)
11. Klöckner, S.: Handbuch Gregorianik. ConBrio, Regensburg (2009)
12. Mazzola, G.: Geometrie der Töne. Birkhäuser, Basel (1990)
13. Reimer, M.: Der Klang als Formel. Oldenbourg, München (2010)
14. Rossing, T.D., Fletcher, N.H.: Principles of Vibration and Sound, 2. Aufl. Springer, New York (2004)
15. Schröder, E.: Mathematik im Reich der Töne. BSB Teubner, Leipzig (1990)
16. Schüffler, K.: Die Tonleiter und ihre Mathematik. Springer. Heidelberg (2022)
17. Schüffler, K.: Pythagoras. der Quintenwolf und das Komma. Springer Fachmedien, Wiesbaden (2017)
18. van der Waerden, B.L.: Geometry and Algebra in Ancient Civilizations. Springer, Berlin (1983)
19. Wright, D.: Mathematics and Music. AMS vol 29. (2009)

Supplementary Literature on the Subject of Mathematics

20. Heuser, H.: Lehrbuch der Analysis. ViewegTeubner, Stuttgart (1991)
21. Hewitt, E., Stromberg, K.: Real and Abstract Analysis. Springer, Heidelberg (1965)
22. Walter, W.: Analysis (Grundwissen). Springer, Heidelberg (1985)

Supplementary Literature on the Organ

23. Ars organi, internationale Zeitschrift für das Orgelwesen, GdO (Hg.)
24. Reichling, A. (Hrsg.): Orgel. Bärenreiter, Kassel (2001)

Index

Printed in the United States
by Baker & Taylor Publisher Services